Advances in Anatomy
Embryology and Cell Biology

Vol. 154

Editors

F. Beck, Melbourne   B. Christ, Freiburg
W. Kriz, Heidelberg   E. Marani, Leiden
R. Putz, München   Y. Sano, Kyoto
T. H. Schiebler, Würzburg
K. Zilles, Düsseldorf

Springer

Berlin
Heidelberg
New York
Barcelona
Budapest
Hong Kong
London
Milan
Paris
Singapore
Tokyo

H. J. ten Donkelaar

Development and Regenerative Capacity of Descending Supraspinal Pathways in Tetrapods: A Comparative Approach

With 45 Figures and 9 Tables

 Springer

H.J. TEN DONKELAAR
Department of Neurology
University Hospital Nijmegen
and The Nijmegen Institute for Neurosciences
University of Nijmegen
P.O. Box 9101
6500 HB Nijmegen, The Netherlands
(e-mail: H.tenDonkelaar@czzorlnm.azn.nl)

ISBN 3-540-66466-1 Springer-Verlag Berlin Heidelberg NewYork

Library of Congress-Cataloging-in-Publication-Data
Donkelaar, H. J. ten (Henrik Jan), 1946– . Development and regenerative
capacity of descending supraspinal pathways in tetrapods: a comparative
approach / H.J. ten Donkelaar. p. cm. – (Advances in anatomy, embryology,
and cell biology; v. 154)
  Includes bibliographical references and index.
  ISBN 3-540-664661 (softcover: alk. paper)
1. Efferent pathways-Regeneration. 2. Nerves-Growth. 3. Xenopus-Develop-
ment. 4. Comparative neurobiology. I. title. II. Series.
QP371 .D665 1999   573.8'5–dc21

Production: PRO-EDIT GmbH, 69126 Heidelberg, Germany
Printed on acid-free paper – SPIN: 10698839   27/3136wg - 5 4 3 2 1 0

Acknowledgements

The research described in this survey was supported by grants from the Faculty of Medical Sciences of the University of Nijmegen, by grants from the Netherlands Organization for the Advancement of Science (NWO), and by a collaborative research grant from NATO. The permission to use figures of, among others, Peter van Mier, Bernd Fritzsch, Alan Roberts, Joel Glover, Agustín González, Jessie Gribnau, Teresa Cabana and George Martin, and the excellent technical support by Roelie de Boer-van Huizen are gratefully acknowledged. The idea for this survey arose during a symposium entitled "Development, plasticity and regeneration of descending supraspinal pathways," held at the European Neuroscience Association Annual Meeting in Vienna (September 6, 1994). I am most grateful to George Martin for his encouragement to write this, hopefully not too dated, survey.

1 Introduction

1.1
Phylogenetic Constancy of Descending Supraspinal Pathways

During the evolution of vertebrates, various locomotor patterns such as swimming, walking, running, jumping, flying and burrowing were developed (ten Donkelaar 1999). Each of these diverse vertebrate locomotor modes is derived from the fundamental swimming pattern, i.e., lateral undulation, present in most aquatic chordates. Paired fins did not evolve until later. When the land vertebrates arose, the lateral paired fins were converted into organs of locomotion, on the ground or in the air. Finally, particularly in primates, modifications of the distal parts of the extremities (usually the forelimbs) allowed the manipulation of the environment.

The descending pathways from the telencephalon and the brain stem to the spinal cord represent the instruments by which the central nervous system (CNS) steers these diverse locomotor modes (see Kuypers 1981; Kuypers and Martin 1982). Direct telencephalospinal pathways such as the mammalian corticospinal tract are generally absent in nonmammalian tetrapods. However, in origin, course and site of termination, descending brainstem pathways show remarkable similarities in amphibians, reptiles, birds and mammals, suggesting a *phylogenetic constancy* of descending input from the brain stem to the spinal cord in vertebrates (ten Donkelaar 1982). In nonmammalian vertebrates, the bulk of the descending supraspinal tracts is formed by reticulospinal pathways. Reticulospinal neurons constitute the most ancient descending system involved in motor control in all vertebrate classes from cyclostomes to mammals (Shapovalov 1972, 1975; Rovainen 1979; ten Donkelaar 1982, 1994; Grillner et al. 1988, 1991; Fetcho 1992). With the appearance of extremities, the development of an adequate neural control system for the steering of limb movements became apparent. It is likely that the rubrospinal tract plays an important role in this mechanism. The relatively simple rubrospinal tract in nonmammalian tetrapods such as limbed amphibians and reptiles (ten Donkelaar 1982, 1988), and the absence of a corticospinal tract are consistent with the more rigid and stereotyped motor behavior of these vertebrates. The greater repertoire of movements in mammals involves a more complex circuitry including an increased development of the corticospinal tract (Kuypers 1981; Armand 1982; Heffner and Masterton 1983; Nudo and Masterton 1990a,b; Nudo et al. 1995; Porter and Lemon 1993; Darian-Smith et al. 1996).

1.2
Attractive Animal Models for Studies
on the Development of Motor Systems

The phylogenetic constancy of descending supraspinal pathways, at least of those arising in the brain stem, probably implies a comparable pattern of development, presumably a *developmental sequence* in the formation of these central motor pathways (ten Donkelaar 1982). Although the zebrafish embryo is increasingly being used for developmental studies, e.g., for molecular and genetic analyses of axon guidance (Kimmel 1993; Schier 1997), for studies on the development of motor systems, anurans such as the clawed toad, *Xenopus laevis*, chicken embryos and opossums (the North American opossum, *Didelphis virginiana*, and the Brazilian gray short-tailed opossum, *Monodelphis domestica*) are also very attractive animals. During their development from fertilized egg to juvenile animal, most anurans are easily accessible for experimental work. *Xenopus laevis* tadpoles are particularly useful for such studies as they are remarkably transparent, so that their CNS is visible in vivo. In the same manner as chicken embryos (Glover et al. 1986; Glover and Petursdottir 1988, 1991), in vitro preparations can be used for tracing studies (Luksch et al. 1996). Marsupials are born in an immature state, many of them having an extended postnatal development (Martin et al. 1978; Tyndale-Biscoe and Janssens 1987; Saunders et al. 1989; Wang et al. 1992), permitting experimentation at stages of development corresponding with those occurring in utero in placental mammals. Moreover, an in vitro approach can be used, i.e., experiments can be carried out in an isolated brain-spinal cord preparation which can be maintained in long-term culture (Nicholls et al. 1990; Saunders et al. 1992; Møllgård et al. 1994). The availability of lipophilic carbocyanine dyes such as DiI (Honig and Hume 1986, 1989) has opened the possibility of studying the development of descending supraspinal pathways in placental animals (Auclair et al. 1993, 1999; Easter et al. 1993; de Boer-van Huizen and ten Donkelaar 1999). In late embryonic and fetal rats, horseradish peroxidase (HRP) was applied via an intrauterine approach by Kudo et al. (1993) and in particular Lakke (Lakke and Marani 1991; Lakke 1997). Lakke's data provided a detailed time-table of development of the descending supraspinal projections in rats, at least for prenatal stages from embryonic day 16 (E16) onwards. Recently, biotinylated dextran amine (BDA) was used in an isolated brain preparation (de Boer-van Huizen and ten Donkelaar 1999).

1.3
Axonal Pathfinding

The development of the CNS proceeds in a series of decisions made by neuroectodermal precursors and their progeny, decisions that progressively outline the fate of a differentiated neuron in terms of its regional identity, position, cell type, pattern of axonal connections, morphology, and neurochemical phenotype (Jacobson 1991; McConnell 1991). Multiple cues may act in concert to guide axons along a given pathway. Harrison (1910) suggested that the earliest axons to grow out, the *pioneer fibers*, function as mechanical guides for secondary fibers that grow out later by forming bundles or fascicles. The concept of pioneer fibers has been supported by evidence that the first axon to grow out almost always reaches its target by the shortest route and without error (Jacobson 1991). This has been observed in *Xenopus laevis* (Jacobson and Huang 1985) and insect embryos (Goodman and Tessier-Lavigne 1997; Nassif et al. 1998).

Several mechanisms have been suggested to guide the pioneer fibers to their targets. Already in 1887, His postulated that the glia or the interstitial spaces between them may form preformed extracellular tunnels. Singer et al. (1979) proposed the blueprint hypothesis which holds that axons grow where an enlarged extracellular space is provided. Evidence for preformed glial pathways that act as guides for axons has been found for the optic nerve (Silver and Sidman 1980), the corpus callosum (Silver et al. 1982, 1993), the regenerating newt tail spinal cord (Egar and Singer 1972; Nordlander and Singer 1978), and the developing *Xenopus laevis* tail cord (Nordlander and Singer 1982a, b). In most regions of the brain, however, there is little direct evidence for the existence of such channels, and growth cones may actively generate spaces in the absence of preformed channels by releasing proteases that modify their immediate environment (Jessell and Dodd 1991). Recent studies on axonal guidance have focussed primarily on the identification of molecules that regulate growth cone extension and navigation (Tessier-Lavigne and Goodman 1996; Goodman and Tessier-Lavigne 1997). Growth cones establish appropriate connections with their targets during development by responding to both positive and negative guidance cues. At least four different mechanisms – contact attraction, chemoattraction, contact repulsion, and chemorepulsion – appear to be involved.

Tract formation has been studied in greatest detail in invertebrates (see Goodman 1996; Nassif et al. 1998) in which it has been shown that a small number of early generated neurons (*pioneer neurons*) play an important role in the establishment of axon tracts. These neurons lay down an *axonal scaffold* containing guidance cues that are available to later generated growth cones. There is now evidence that similar mechanisms may operate in the vertebrate CNS as has been shown for the zebrafish, *Danio rerio* (Chitnis and Kuwada 1990; Wilson et al. 1990; Ross et al. 1992; Kimmel 1993), lampreys (Kuratani et al. 1998), *Xenopus laevis* (Nordlander 1987; Hartenstein 1993; Easter et al. 1994), birds (Chédotal et al. 1995), and mammals (Easter et al. 1993). Several markers such as acetylcholinesterase, neurofilament proteins, HNK-1 and anti-acetylated tubulin antibodies have been used to stain early phases of neuronal development and axonal outgrowth (e.g., Moody and Stein 1988; Hanneman and Westerfield 1989; Szaro et al. 1989, 1991; Nordlander 1989, 1993; Chitnis and Kuwada 1990; Metcalfe et al. 1990; Hartenstein 1993; Moody et al. 1996).

1.4
Regenerative Capacity of Descending Supraspinal Pathways

The regenerative capacity of descending supraspinal pathways in terrestrial vertebrates greatly differs (Eidelberg 1981). Functional recovery occurs after transection of the spinal cord of urodeles (e.g., Stensaas 1983; Davis et al. 1989a, 1990). Following injury of the salamander spinal cord, the ependymal cells surrounding the central canal retain their viability, and form a new neural tube which is subsequently invaded by descending axons (Egar and Singer 1972; Nordlander and Singer 1978). The functional recovery is linked to the developmental stage of the system or region that is lesioned (Holder and Clarke 1988; Oorschot and Jones 1990). In contrast to urodeles, in anurans no significant regeneration of descending supraspinal pathways occurs after transection of the spinal cord. In tadpoles subjected to cord transection, however, spinal cord continuity is readily restored (e.g., Beattie et al. 1990). The chicken embryo is also capable of axonal repair after complete transection of the spinal cord, but successful recovery appears not to exceed embryonic day 14 (Shimizu et al. 1990; Hasan et al. 1991, 1993).

In newborn opossums, repair of spinal cord connections occurs rapidly (Martin et al. 1994; Nicholls et al. 1994; Nicholls and Saunders 1996; Martin and Wang 1997; Saunders 1997). But, as the CNS matures, the capacity for regeneration ceases abruptly. In adult opossums, as in other mammals (see Schwab and Bartholdi 1996), the spinal cord does not regenerate. The critical period at which regeneration stops coincides with myelination and oligodendrocyte development. CNS myelin is inhibitory to the regeneration of injured CNS axons (Schwab et al. 1993). Blocking of myelin-associated inhibitors of axonal growth extends the period during which axonal regeneration is possible (Schnell and Schwab 1990; Keirstead et al. 1995).

1.5
Scope of the Present Review

In the present survey, following some introductory notes on the organization of descending supraspinal pathways, and on systems for staging tetrapod embryos, current knowledge on the neurogenesis, axonal outgrowth, synaptogenesis, and developmental plasticity of the central motor pathways in tetrapods including the sparse data available for man, will be discussed. These data will be placed in the perspective of the development of the spinal cord and, where possible, correlated with functional data. Emphasis will be on studies in the clawed toad, *Xenopus laevis*, chicken embryos, and opossum and rodent data. It will be shown that: (1) the outgrowth of descending supraspinal pathways is the result of a coordinated program; (2) the pattern of early descending axonal tracts is similar in all vertebrate groups; (3) the formation of descending supraspinal pathways occurs according to a developmental sequence; (4) the earliest descending supraspinal fibers arrive in a rather immature spinal cord; and (5) the regenerative capacity of descending supraspinal pathways depends on the developmental stage in which the particular pathways arise.

2 Materials and Techniques

The core of the present survey is formed by the experimental data obtained in *Xenopus laevis*. The description of the techniques used is therefore largely restricted to the procedures used for this anuran. The *X. laevis* larvae were obtained by induced breeding (intraperitoneal Pregnyl injection), and kept in fresh tap water at room temperature (20–22°C). A wealth of data on various aspects of the circuitry of both tadpoles and adult *X. laevis* were available, obtained with tract-tracing techniques and immunohistochemical procedures.

2.1
Cytoarchitectonic Analysis

For the developmental stages of *Xenopus laevis* studied, paraffin embedded transversely sectioned (7–15 μm depending on the stage) brains and spinal cords were available. Adjacent sections were stained with cresylecht violet, with silver proteinate according to Bodian (1936) and, in older stages, according to Klüver and Barrera (1953). Moreover, material stained with Rager's modification of the Bodian technique (Rager et al. 1979) was available.

2.2
Immunohistochemical Procedures

The present survey includes immunohistochemical data on the development of monoaminergic pathways. The development of raphespinal projections was studied in *Xenopus laevis* larvae ranging from stage 25 to juveniles (stage 66). Before fixation larvae were deeply anesthetized in tap water containing an overdose of MS 222 (Sandoz; 50 mg/100 cc). Fixation was done by immersion (stages 25–48) or by intracardial perfusion (older stages) with a mixture of 4% paraformaldehyde, 0.05% glutaraldehyde and 0.2% picric acid in 0.1 M phosphate buffer (final pH 7.2–7.4). After the initial fixation at room temperature the CNS was dissected out, further fixed for another 6 h at 4°C after which the meninges were removed, washed several times in 0.1 M phosphate-buffered saline (PBS, pH 7.2), cryoprotected with sucrose, and cut on a cryostat or a freezing microtome. Sections were either incubated according to the indirect immunofluorescence technique and labeled with fluorescein isothiocyanate (FITC) or with the peroxidase-antiperoxidase immunohistochemical procedure. The preparation and specificity of the antibody against serotonin used were described by

Steinbusch et al. (1983). For details on the procedures see van Mier et al. (1986). For the description of the techniques for visualizing tyrosine hydroxylase and dopamine-immunoreactivity, data on which are included in this review, the reader should consult González et al. (1994a,b).

2.3
Tract-Tracing Studies

Modern tract-tracing techniques can readily be applied to the developing CNS, both in vivo and in vitro (ten Donkelaar and Nicholson 1998). In particular horseradish peroxidase (HRP), fluorescent tracers such as Fast Blue and Nuclear Yellow, fluorescent (FDA, RDA) and biotinylated (BDA) dextran amines, and lipophilic carbocyanine tracers such as DiI are widely used as neuronal tracers. For most tracing experiments in *Xenopus laevis* discussed, HRP (Boehringer or Serva) was used. The HRP experiments were carried out under surgical anesthesia (MS 222, Sandoz) with a Zeiss binocular operation microscope. In early stages, the skin was pierced with a very thin needle, and after crushing the spinal cord, HRP was applied as dry crystals. In older stages, following a midline skin incision and separation of the dorsal musculature, a small HRP slow-release gel was implanted into the spinal cord with a fine-tipped forceps. After survival times of 1–3 days, the tadpoles, under MS 222 anesthesia, were fixed by immersion in a mixture of 1.25% glutaraldehyde and 1% formaldehyde in 0.1 M phosphate buffer (pH 7.4) or perfused through the heart with a similar mixture. After 2 h, the brain and spinal cord were dissected out, postfixed in the same solution for another hour, and placed in PBS containing 30% sucrose at 4°C for 1–6 days. The brain and spinal cord were further processed either as wholemounts and/or transverse sections. Preparations to study as wholemounts were stained according to a modification of the heavy metal intensification of the 3,3-diaminobenzidine tetrahydrochloride (DAB) reaction introduced by Adams (1981). The stained preparations were dehydrated in alcohol series, cleared in cedarwood oil and studied as wholemounts. Because of their size, the CNS of larvae between stages 40 and 48 were pretreated with 0.3% Triton X-100. After studying and photography, some of the wholemounts were processed via acetone, alcohol, methylbenzoate and acetylsalicylate for embedding in Paraplast and cut into transverse sections of 5–10 μm. Older stages were studied as transverse sections only.

For sectioning, after fixation, several preparations were embedded in a gelatine-sucrose mixture (30% sucrose in PBS with 15% gelatine). About two to six preparations were embedded in one gelatine block and stored overnight in 4% formaldehyde. The material was subsequently cut into transverse sections of 20 μm on a freezing microtome. Sections were either incubated according to Adams' (1981) heavy metal intensification of the DAB technique or according to Mesulam's (1978) tetramethylbenzidine (TMB) technique. In some cases, the larger brains were embedded in polyacrylamide and sectioned on a vibratome at 100 μm. For electron microscopy some brains were osmicated and embedded in Epon.

In some experiments various 10 or 3 kD lysine-fixable dextran amines conjugated to the fluorochromes tetramethylrhodamine (RDA, D-1817 and D-3308; Molecular Probes, Eugene, Ore.) and fluorescein (FDA, D-1820 and D-3306) were used. RDA and FDA were applied recrystallized from distilled water onto fine tungsten needles. After survival times of 2–3 days, the animals were re-anesthetized with an overdose of MS 222, and perfused through the heart with 0.1 M phosphate buffer (pH 7.4) followed by a fixative containing 4% paraformaldehyde in phosphate buffer. The brain and spinal cord were dissected out, embedded in polyacrylamide, left overnight in 15% sucrose in 0.1 M phosphate buffer, and cut on a freezing microtome at 40 μm. They were mounted in glycerin-gelatin and viewed with a Zeiss fluorescence microscope with appropriate filter combinations. FDA and RDA were also applied in studies on the plasticity and regeneration of descending supraspinal pathways, requiring much longer survival times (see Sects. 5.4.1 and 5.4.3).

In studies on the early development of descending pathways in rat embryos, the carbocyanine dye 1,1'-dioctadecyl-3,3,3',3'-tetramethylindocarbocyanine per-chlorate (DiI) was applied in fixed rat embryos 11–14 gestational days of age (E11–E14), whereas biotinylated dextran amine (BDA) was used in isolated rat embryos (de Boer-van Huizen and ten Donkelaar 1999). E11–E14 embryos were fixed for a few days in 4% paraformaldehyde in phosphate buffer (pH 7.4). A small DiI crystal (Molecular Probes, D-282) was applied to the rostral spinal cord, just behind the obex. After that the embryos were kept in the same fixative for a period of 4–6 weeks at 37°C. Subsequently, the embryos were rinsed in phosphate buffer, kept overnight in 15% sucrose in phosphate buffer, and embedded in polyacrylamide (de Boer-van Huizen 1989). The embryos were cut on a freezing microtome (MICROM HM 440 E) into 40 μm thick sections and mounted in glycerin-gelatin. The sections were studied with a Zeiss Axiovert 35 M fluorescence microscope using a rhodamine filter (550–560 nm).

Isolated rat embryos (E10-E12) were immersed in an iced Ringer solution (Karamian et al. 1991) that had been oxygenated with carbogen (95% O_2, 5% CO_2). The rostral spinal cord was isolated and a small BDA crystal (3 kD BDA, Molecular Probes, D-7135) was placed just behind the obex. After the application the embryos were placed in a perfusion chamber (see Luksch et al. 1996) and superfused overnight with freshly oxygenated Ringer solution at room temperature. The next day, the embryos were fixed with 4% paraformaldehyde in phosphate buffer (pH 7.4), embedded in polyacrylamide or in Epon if treated as wholemounts, and cut coronally or sagittally on a vibratome. In the 40 μm thick sections, BDA was visualized with an avidin-biotin complex (Vectastain).

2.4
[^{3}H]-Thymidine Studies

To determine the time of origin of neurons in the brain stem and spinal cord, *Xenopus laevis* embryos and larvae between stages 10 and 50 were injected with [^{3}H]-thymidine (van Mier 1986). In embryos up to stage 40, a volume of 0.02 μl containing 0.1 μCi [^{3}H]-thymidine (Radiochemical Centre, Amersham, U.K.; specific activity 75 Ci/mmole) was injected into the yolk sack with a glass micropipette. Older stages were anesthetized with MS 222 before the injection after which the abdominal cavity was opened with fine tungsten needles. After a single injection of 0.1–0.7 μCi [^{3}H]-thymidine in a volume of 0.02–0.1 μl the embryos and larvae were transferred to small vials containing limited amounts of water and left for 1 h. After the injection procedure the animals were placed in small water tanks and kept at 20–22°C. Most of the embryos were allowed to complete development through metamorphosis until three

7

weeks after stage 66. Several larvae which received a [3H]-thymidine injection at stages 28, 32, 37/38, 40 and 50 were also used for double labeling experiments in which HRP was used as a retrograde tracer (modified after Nowakowski et al. 1975).

The animals which had only received a [3H]-thymidine injection were heavily anesthetized in MS 222 solution (100 mg/ml water), and perfused transcardially with 0.1 M PBS followed by Bouin's fixative. After perfusion the specimens were left in fresh Bouin for 6 h after which the CNS was removed. Animals of the combined [3H]-thymidine/HRP experiments were, under deep anesthesia, perfused with PBS followed by an ice-cold mixture of 2.5% glutaraldehyde and 1.0% formaldehyde in PBS (pH 7.2). Then the CNS was removed and stained as a whole for the presence of HRP labeled cells as described in Sect. 2.3.

After staining, all specimens were washed in fresh PBS, dehydrated through graded alcohol series, transferred to methylbenzoate and amylacetate, and finally, after embedding in Paraplast, transverse sections (7 µm) were cut serially and mounted on coated glass slides. After dewaxing, the glass slides were dipped (Berry and Rogers 1965) in Ilford Nuclear Emulsion at 37°C and dried in the air for at least 4 h. The slides were then kept in lead boxes for 2–3 weeks. The sections were developed in amidol developer at 15°C for 10 min, fixed in 30% sodium thiosulfate and counterstained with neutral red or cresylecht violet.

3 Descending Pathways to the Spinal Cord in Tetrapods: A Brief Outline

Available experimental data on the cells of origin, funicular trajectory and site of termination of descending supraspinal pathways in amphibians (ten Donkelaar et al. 1981; ten Donkelaar 1982; Tóth et al. 1985; Naujoks-Manteuffel and Manteuffel 1988), reptiles (ten Donkelaar et al. 1980; ten Donkelaar 1982), birds (Cabot et al. 1982; Gross and Oppenheim 1985; Webster and Steeves 1988; Webster et al. 1990) and mammals (see Kuypers 1981 for review; for data in opossums see Martin et al. 1975; Crutcher et al. 1978; Holst et al. 1991) show that throughout tetrapods a basic pattern in the organization of descending pathways is present. The most notable difference between nonmammalian tetrapods and mammals is the apparent absence of somatomotor cortical areas giving rise to long descending projections to the spinal cord.

In Fig. 1, these experimental data are summarized for the clawed toad, *Xenopus laevis*, the pigeon *Columba livia*, and the North American opossum, *Didelphis virginiana*. As regards the course and site of termination of the descending pathways from the brain stem to the spinal cord, a classification can be made as advocated in mammals (Lawrence and Kuypers 1968a,b; Kuypers 1981) into *lateral* and *medial* systems. Interstitiospinal, reticulospinal and vestibulospinal pathways pass via the ventral funiculus and ventral parts of the lateral funiculus, and terminate in the mediodorsal parts of the ventral horn and the adjacent parts of the intermediate zone. This medial system is functionally related to postural activities and progression, and constitutes a basic system by which the brain exerts control over movements. The lateral system is composed of rubrospinal fibers, certain reticulospinal fibers, and raphespinal fibers arising in a rostral, magnocellular part of the medullary raphe nucleus. The rubrospinal tract terminates in lateral and dorsal parts of the intermediate zone. In mammals, the rubrospinal tract, at least in regard to the extremities, superimposes on the general motor control, via the medial system, the capacity for the independent use of the extremities, particularly of the hand (Lawrence and Kuypers 1968a,b; Holstege and Kuypers 1982; Gibson et al. 1985; McCurdy et al. 1987; Cheney et al. 1991). The rubrospinal tract shows a greater somatotopic organization in more dextrous mammalian species such as monkeys (Castiglioni et al. 1978; Larsen and Yumiya 1980) than in less dextrous mammals such as opossums (Martin et al. 1981). The human rubrospinal tract is indistinct (Nathan and Smith 1981). For nonmammalian tetrapods it is likely that a rubrospinal tract is related to the presence of limbs (ten Donkelaar 1988).

In mammals, a direct corticospinal tract gradually supersedes the descending projections from the brain stem. The *corticospinal tract* invariably arises from layer V

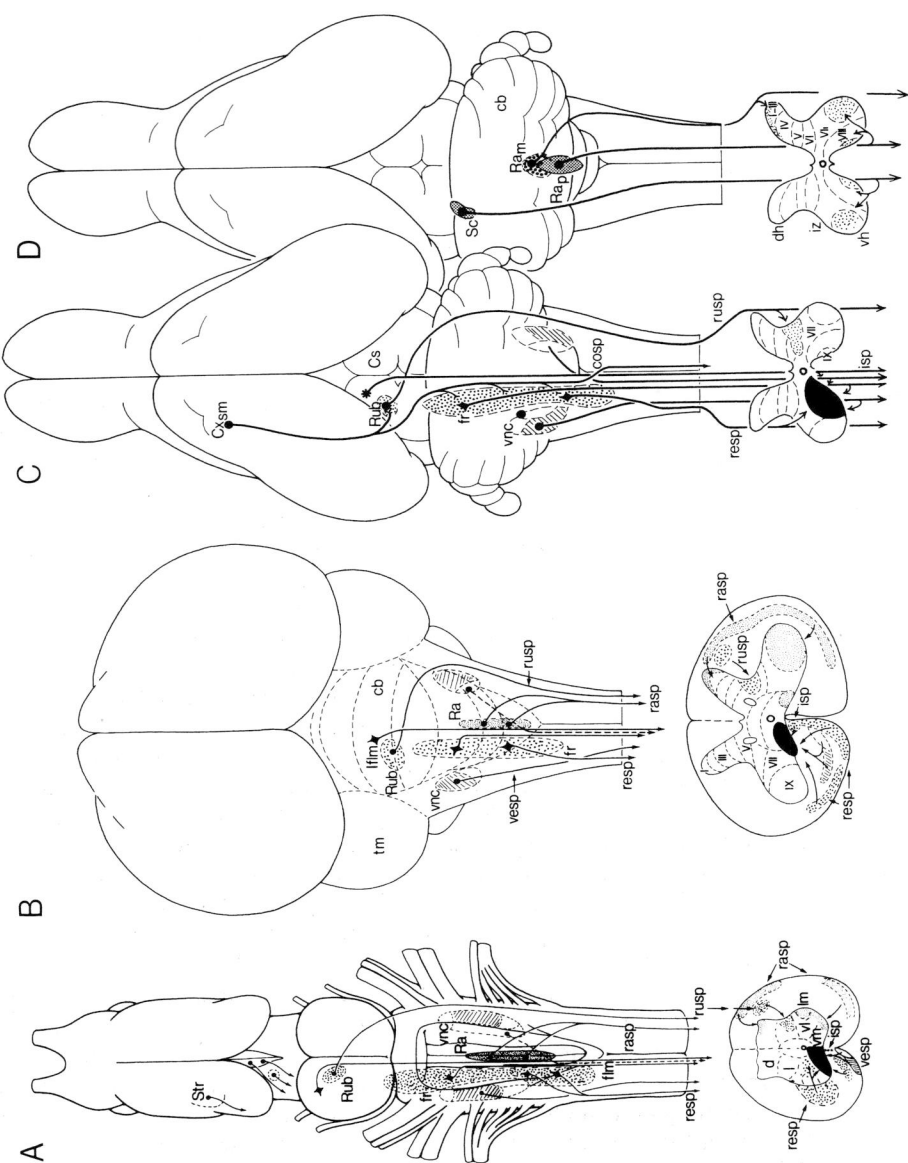

Fig. 1. Summary of experimental data on descending supraspinal pathways in: **A** the clawed toad, *Xenopus laevis* **B** the pigeon, *Columba livia* and **C,D** the North American opossum, *Didelphis virginiana*. *cb*, cerebellum; *cosp*, corticospinal tract; *Cs*, colliculus superior; *Cx*ₛₘ, somatomotor cortex; *d, l, lm, vl, vm*, dorsal, lateral, lateral motor, ventrolateral and ventromedial fields of the spinal gray matter; *dh*, dorsal horn; *flm*, fasciculus longitudinalis medialis; *fr*, formatio reticularis; *Iflm*, interstitial nucleus of the flm; *isp*, interstitiospinal tract; *iz*, intermediate zone; *Ra*, raphe nucleus; *Ra*ₘ, nucleus raphes magnus; *Ra*ₚ, nucleus raphes pallidus; *rasp*, raphespinal projections; *resp*, reticulospinal projections; *Rub*, nucleus ruber; *rusp*, rubrospinal tract; *Sc*, subcoeruleus area; *Str*, striatum; *tm*, tectum mesencephali; *vesp*, vestibulospinal projections; *vh*, ventral horn; *vnc*, vestibular nuclear complex; *I-X*, laminar subdivision of spinal gray matter. (After ten Donkelaar 1982, 1999)

10

pyramidal cells, particularly from rostral, frontal parts of the cerebral cortex (Fig. 2). Both motor and somatosensory cortices give rise to corticospinal projections (see Kuypers 1981; Armand 1982; Nudo and Masterton 1990a, b). The corticospinal projection is predominantly contralateral, but in many insectivores (e.g., moles) largely ipsilateral. Moles have dense corticospinal projections from the forelimb representation of the somatosensory and motor cortices which may reflect sensorimotor specializations related to digging (Catania and Kaas 1997). In the North American opossum, the motor and somatosensory cortices overlap, and corticospinal projections are restricted to the cervical and upper thoracic cord (Martin and Fisher 1968). They pass by way of the dorsal funiculus and terminate predominantly in the medial part of the dorsal horn. With regard to the main terminal distribution of corticospinal fibers, Kuypers (1981) divided the various mammalian species studied into four groups that show an increasingly wider distribution area of cortical fibers in the spinal gray: (1) mammals with corticospinal fibers extending *only* to cervical or midthoracic segments and terminating predominantly in the *dorsal horn* (opossums, tree shrews, rabbits); (2) mammals with corticospinal fibers extending *throughout* the spinal cord and terminating largely in the *dorsal horn* and the *intermediate zone* (rats, most carnivores); (3) mammals with corticospinal fibers extending throughout the spinal cord and terminating in the *dorsal horn*, the *intermediate zone* and *dorsolateral* parts of the *lateral* motoneuronal cell groups (raccoons, most primates); and (4) mammals with corticospinal fibers extending throughout the spinal cord and terminating in the *dorsal horn*, the *intermediate zone* and both *dorsolateral* and *ventromedial* parts of the *lateral* motoneuronal cell groups (apes such as chimpanzees, man). The extent and the site of termination of the corticospinal tract most closely corresponds with dexterity (Heffner and Masterton 1983). Originally, direct *corticomotoneuronal* connections to motoneurons innervating hand and finger muscles were only found in primates (see Kuypers 1981), and a few carnivores such as raccoons and kinkajous (Petras 1969; Wirth et al. 1974). The application of WGA-HRP and other powerful tracers in studies on corticospinal projections in the North American opossum (Martin and Cabana 1985), rats (Liang et al. 1991), cats (Cheema et al. 1984; Armand et al. 1985) and monkeys (Cheema et al. 1984; Ralston and Ralston 1985), however, did not only show distinct projections to the superficial layers of the dorsal horn, but in opossums, rats and cats – as in monkeys – a more ventral extent of corticospinal projections including sparse projections to the lateral motoneuron column. Rats also exhibit a considerable skill in performing finely adjusted distal movements (Kolb and Whishaw 1983; Whishaw and Kolb 1988), as do hamsters (Kalil and Schneider 1975). Direct corticomotoneuronal projections are not restricted to forelimb motoneurons. In many monkeys, the motor cortex also projects to hindlimb motoneurons, and, moreover, in spider (*Ateles*) and woolly (*Lagothrix*) monkeys, to motoneurons innervating the muscles of the prehensile tail (Petras 1968).

The bulk of descending supraspinal pathways in nonmammalian tetrapods is formed by *reticulospinal fibers*. Forebrain structures exert their influence on spinal motor mechanisms via the brainstem reticular formation (ten Donkelaar 1990). Retrograde tracer studies in nonmammalian tetrapods (ten Donkelaar et al. 1980, 1981, 1987; Cabot et al. 1982; Newman et al. 1983; Gross and Oppenheim 1985; Tóth et al. 1985; Webster and Steeves 1988; Webster et al. 1990) showed a much more extensive reticulospinal projection than found with the classical degeneration techniques (ten Donkelaar 1982), in keeping with data in mammals (e.g., Crutcher et al. 1978; Tohyama

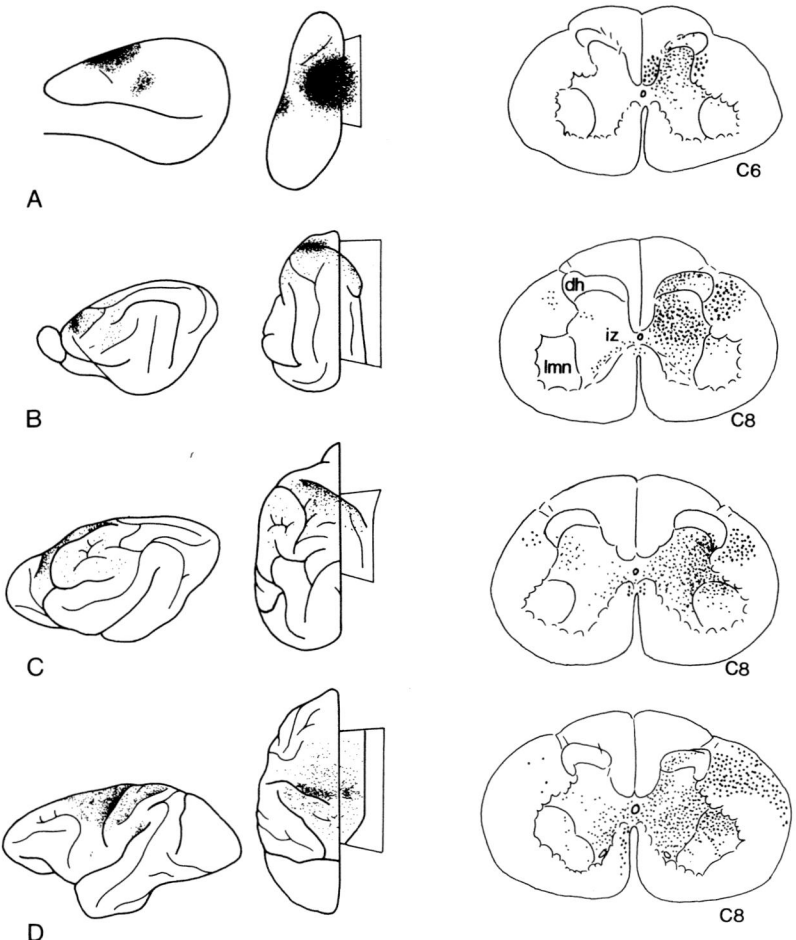

Fig. 2. The distribution of the cells of origin of the corticospinal tract shown in lateral and dorsal views of the brains of: **A** the North American opossum, *Didelphis virginiana* **B** the domestic cat, *Felis catus* **C** the raccoon, *Procyon lotor* **D** the rhesus monkey, *Macaca mulatta* (after Nudo and Masterton 1990a). The mirrors in the dorsal views show the medial extent of corticospinal neurons on the medial side of the hemisphere. Also shown are the sites of termination of the corticospinal tract in the cervical enlargement of the spinal cord. *dh*, dorsal horn; *iz*, intermediate zone; *lmn*, lateral motoneuron column. (**A** after Martin and Cabana 1985; **B** after Buxton and Goodman 1967 as well as Petras 1969; **C** after Cheema et al. 1984; **D** after Kuypers 1981 and Cheema et al. 1984)

et al. 1979a,b; Nudo and Masterton 1988; Holst et al. 1991). Anterograde tracer studies (e.g., Cabot et al. 1982; Holstege and Kuypers 1982; Martin et al. 1982a,b; Tan and Miletic 1990) demonstrated a much wider funicular distribution of reticulospinal and raphespinal projections, and terminal fields not only in the intermediate zone, but also in the superficial layers of the dorsal horn and in the motoneuron area. These anterograde tracing studies complement immunohistochemical studies on the distribution of noradrenergic (e.g., Björklund and Skagerberg 1982; Pindzola et al. 1988; González and Smeets 1991; González et al. 1993), and serotonergic (e.g., Steinbusch 1981; Björklund and Skagerberg 1982; Tan and Miletic 1990; Okado et al. 1992) fibers in the spinal cord. Kuypers (1982; Holstege and Kuypers 1987) considered the extensive noradrenergic (from the subcoeruleus area) and serotonergic (from the nucleus raphes pallidus) brainstem projections to the somatic motoneuronal cell groups as a *third* component of the motor system. This third component of the descending supraspinal control system may be instrumental in providing motivational drive in the execution of movements and play a role in fight and flight. The brainstem areas in question receive an important projection from several components of the limbic system (Holstege and Kuypers 1987; Holstege 1991). This implies that the "emotional brain" can exert a powerful influence on all regions of the spinal cord. Holstege (1991, 1992) introduced the name "*emotional motor system*" for the forebrain and brainstem structures involved.

4 Staging Systems

Differences between the brains of species belonging to different vertebrate classes are evident from the onset of neurulation (Richardson et al. 1997). Nevertheless, embryos of different vertebrate classes share a common organization and at, for instance, the tailbud stage, have many features in common such as the presence of somites, the neural tube, optic anlagen, the notochord and pharyngeal arches. But, evolution has produced a number of changes in the embryonic stages of vertebrates including differences in body plan, changes in the number of units in repeating series such as somites and pharyngeal arches, and changes in the pattern of growth and the timing of development of different fields (see Richardson et al. 1997). To be able to compare data obtained from studies on embryos, systems for staging prenatal development are advocated.

4.1
Systems for Staging Amphibian Embryos

Various systems of staging amphibian embryos are used. In urodeles, staging systems are available for the spotted salamander, *Ambystoma maculatum* (or *punctatum*; Harrison 1969), the axolotl, *A. mexicanum* (Schreckenberg and Jacobson 1975; Bordzilovskaya et al. 1989), the ribbed newt, *Pleurodeles waltl* (Gallien and Durocher 1957), and for a European salamander, *Triturus helveticus* (Gallien and Bidaud 1959). The development of *A. maculatum* is summarized in Fig. 3. Stage 1 is the unsegmented egg, stages 2–7 are cleavage stages, stages 8 and 9 early and late blastula. Neurulation occurs during stages 13–20 with the formation of the neural groove and plate, the gradual rising of the neural folds, and the closure of the neural folds at stage 20. During stages 21–29, rapid growth of the head process takes place with gradual extension beyond the ventral surface of the embryo, enlargement of the optic vesicles, the appearance of nasal pits and ear vesicles, and of placodes for the lateral line organs. During stages 30–35, a gradual straightening of the head curvature with elongation of the trunk, and progressive lengthening of the tailbud takes place. During stages 36–40, the gills and balancers develop, and the forelimb bud takes on a paddle shape. Further development of the forelimbs, appearance of hindlimb buds, development of the digestive system, and absorption of the yolk occur during stages 41–46. Hatching occurs at stage 46.

For the clawed toad, *Xenopus laevis*, Nieuwkoop and Faber (1967) made their well-known table of development, which is summarized in Fig. 4. The development from fertilized egg to juvenile toad occurs during two months (depending on the

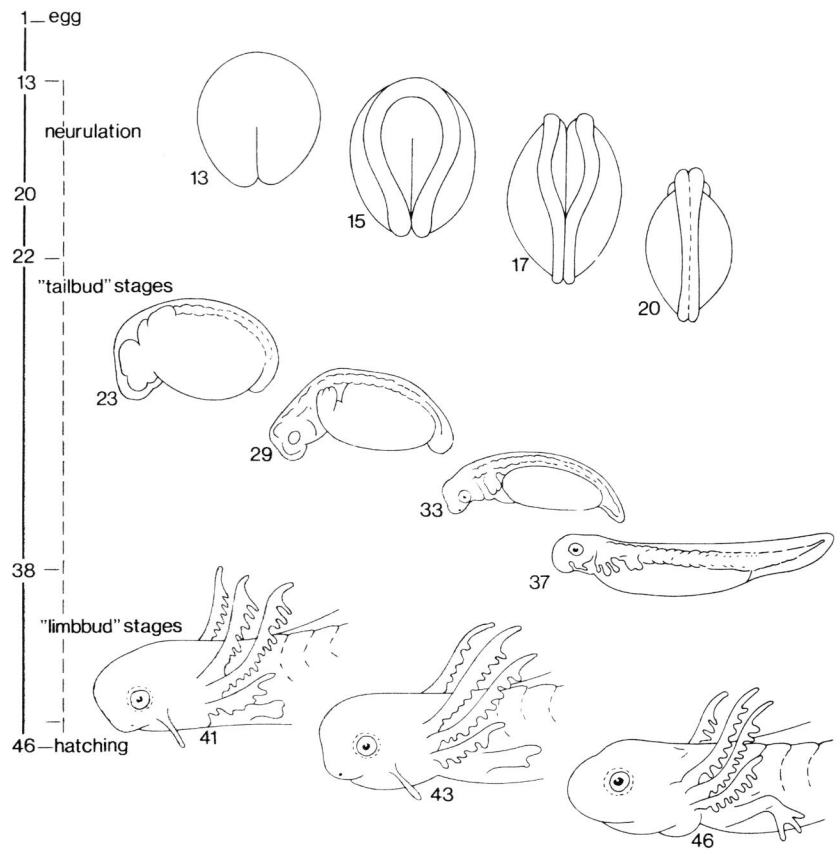

Fig. 3. Developmental stages in *Ambystoma maculatum*. (After ten Donkelaar 1998a; based on Harrison 1969)

ambient water temperature), after which *X. laevis* can reach ages of ten years or more (Deuchar 1975). Its development has been divided into 66 stages. The fertilized eggs pass quickly through the blastula, gastrula and neurula stages into young embryos which hatch around stage 35/38. They start to feed around stage 40 when the yolk sac has been consumed. During metamorphosis, when the tail and gills disappear and limbs and lungs develop, roughly three, overlapping, phases of locomotion can be distinguished (Muntz 1964, 1975; Macklin and Wojtkowski 1973; van Mier et al. 1989): (a) the *immobile phase* (stages 1–24); (b) the *phase of tail locomotion* (approximately stages 20–60), including non-motile (stages 20–22), pre-motile (stages 22–24), early flexure (stages 24–28), early swimming (stages 28–33) and free swimming (from stage 33 on) phases; and (c) the *phase of limb locomotion* (approximately stages 46–66), including non-motile (stages 46–53), pre-motile (stages 53–55), motile (stages 55–59) and fully functional (from stage 59 onwards) stages. The development of the tail starts with the segregation of the first myotomes at stage 17 (Nieuwkoop and Faber 1967; Muntz 1964, 1975; Macklin and Wojtkowski 1973; Elsdale and Davidson 1983; van

16

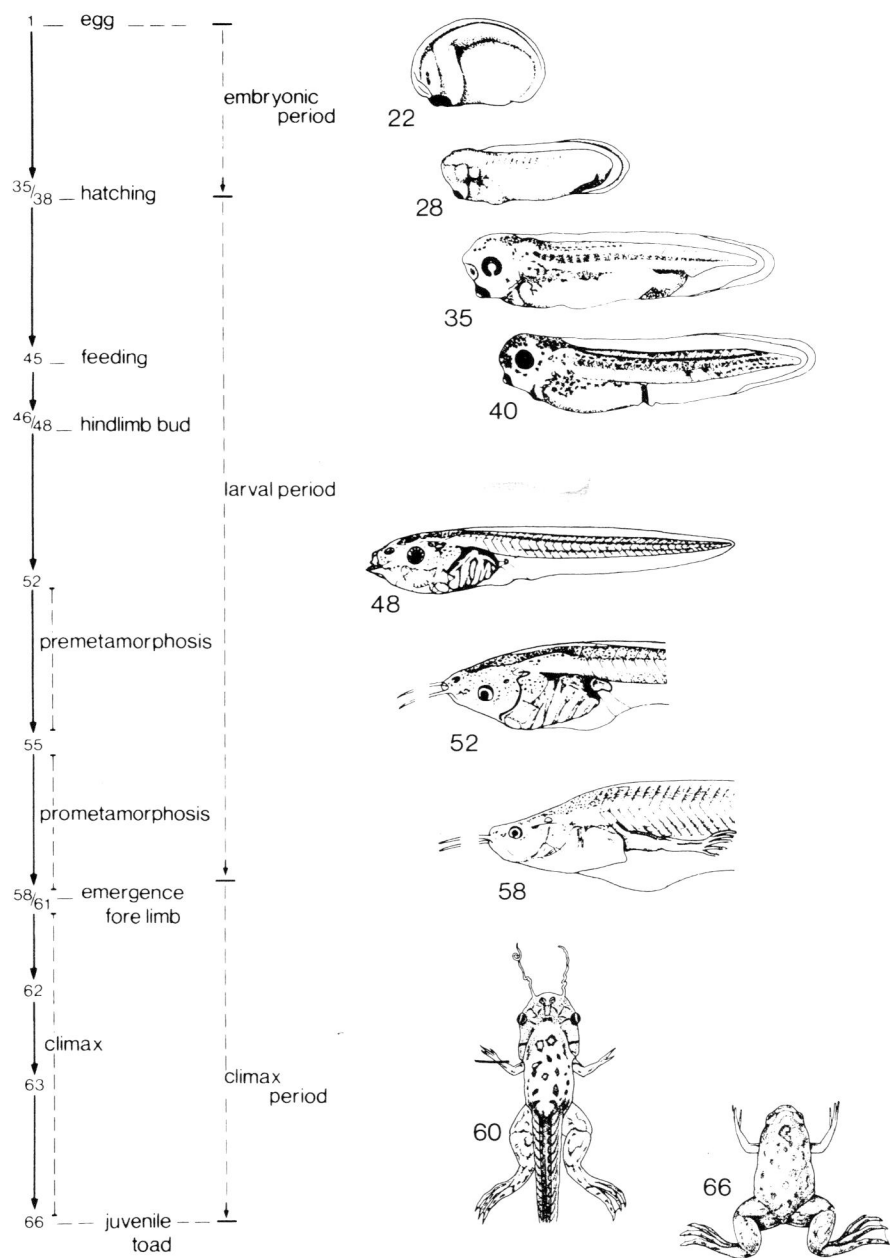

Fig. 4. Developmental stages in the clawed toad, *Xenopus laevis*. (After ten Donkelaar 1998b; based on Nieuwkoop and Faber 1967, and van Mier 1986)

Mier et al. 1989). At stage 25, some 15 consecutive myotomes are present on both sides of the spinal cord. From stage 24 on, short jerky movements can be observed which gradually change to more regular movements at stage 28 (Muntz 1964; van Mier et al. 1989). Between stages 30 and 33 rhythmicity appears in locomotion and the larvae become free swimming, i.e., undulatory waves of bending of the body which pass alternately down either side of the body, can be observed (Kahn et al. 1982; van Mier et al. 1989). The hindlimbs begin to develop at stage 46 (Hughes and Prestige 1967; Nieuwkoop and Faber 1967). Although at stage 54 the first movements of the hindlimbs appear (Hughes and Prestige 1967), the hindlimbs are not used for locomotion until stage 58. From stage 63 on, the tail is immobile, locomotor movements are made with the hindlimbs only, in the bilateral synchronous rhythm so characteristic for swimming of adult anurans. Forelimb buds are visible at stage 48, and emerge at stage 58.

Ranid frogs are usually classified according to the criteria of Taylor and Kollros (1946) for *Rana pipiens*, the common North American frog. Their system for larval stages follows the 25 embryonic stages of Shumway (1940), and is mainly based on hindlimb morphology, beginning with the onset of independent feeding (Table 1). Stages I-V are limb bud stages; foot paddle stages extend from VI to X. During the premetamorphic stages (XI-XVII), the hindlimb becomes progressively larger and of more use to the animal for swimming (see Stehouwer and Farel 1984; Stehouwer 1992). Stages XVIII-XXV, the metamorphic stages, begin with degeneration of the cloacal tailpiece. The forelimbs emerge at stage XX, and the tail undergoes progressive degeneration until it is completely resorbed at stage XXV. Gosner (1960) presented another system for staging the development of *Rana pipiens*, and Manelli and Margaritora (1961) for the green frog, *Rana esculenta* (Table 1).

4.2
Staging Chicken Embryos

Hamburger and Hamilton's (1951) table of normal development of the chicken embryo comprises 46 stages, and includes the period after appearance of the primitive streak (their stage 2) until the stage of the newly hatched chicken (20–21 days). They used morphological criteria: the number of somites to define the early stages 7–14, and after stage 14 (possessing 22 somites), the development of the limbs (Fig. 5). Stage 1, the pre-streak stage was subdivided into 14 (I-XIV) pre-primitive streak stages (Eyal-Giladi and Kochav 1976). Stages HH7-14 are somite stages. At HH7 (23–26 h of incubation) one somite is present and the neural folds are visible in the head region. At the four somite stage HH8 (26–29 h), the neural folds meet at the level of the midbrain. This process extends in the rostral and caudal region. At stage HH9 (after 27 h), closure of the neural tube begins. The enlargement of the cephalic part of the neural tube starts around stage 9–10. From this stage on a pattern of neuromeres develops in the developing brain stem and prosencephalon (Vaage 1969; Keynes and Lumsden 1990; Lumsden 1990; Guthrie 1995; Nieuwenhuys 1998). Neuromeres are transient repetitive bulges in the brain, especially prominent at the hindbrain level. Eight rhombomeres can be distinguished (Lumsden and Keynes 1989; Lumsden 1990), one mesomere in the mesencephalon, and six prosomeres in the forebrain (Puelles et al. 1987).

Table 1. Comparison of timetables for anuran development

Main embryonic events	Xenopus laevis[b]	Rana pipiens[c,d]	Rana esculenta[a]
Embryonic period		Embryonic stages[c]	
Fertilized egg	Stage 1	Stages 1+2	Stage 1
Period of cell cleavage	Stages 2–6	Stages 3–7	Stages 2–6
Blastula stages	Stages 7–9	Stages 8, 9	Stage 7
Early gastrula stages	Stages 10–13	Stages 10–13	Stages 8–12
Neural plate stage	Stage 14	~ Stage 14	~ Stage 13/14
Neural groove stage	Stage 18	~ Stage 15	~ Stages 15–17
Neural folds fused	Stage 20		~ Stage 19
Distinct protrusion of eyes	Stage 22	~ Stage 16	~ Stage 21
Ear vesicles protruding	Stage 26		
Tail bud distinct	Stage 29/30	Stage 18	~ Stage 22/23
Hatching starts	Stage 35/36	~ Stage 20	~ Stage 30/31
Larval period		Larval stages[d]	
Limb bud stages		Stages I–V	
– Hindlimb bud visible	Stage 46	Stage I	Stage 32
– Forelimb bud visible	Stage 48	~ Stage II	
Premetamorphic stages (foot paddle stages)	Stages 52–55	Stages VI–X	Stages 35/36–39
Prometamorphic stages – Emergence of forelimbs	Stages 55–58/61 Stage 58	Stages XI–XVII	Stages 39–44/45
Metamorphic stages	Stages 58/61–66	Stages XVIII–XXV	Stages 44/45–49/50

[a] Manelli and Margaritora 1961.
[b] Nieuwkoop and Faber 1967.
[c] Shumway 1949.
[d] Taylor and Kollros 1946.

Limb primordia can first be recognized at HH15 (50–55 h), but limb buds are not clearly visible before HH17 (52–64 h; see Fig. 5). At HH18 (3 days, E3), leg buds are slightly larger than wing buds, at HH21 (E3.5), this difference is more distinct, and at HH25 (E4.5–5), elbow and knee joints are distinct with a distinct digital plate in the wing, but no digits are visible. At HH26 (E5), the first three toes are distinct. At HH30 (E6.5–7), the three segments of wings and legs are clearly demarcated. The further development of the limbs continues until hatching (HH46, E20–21). Spontaneous movements begin at E3.5–4 and continue up to the time of hatching (Bekoff 1992). Flexures of the trunk appear first, and limb movements appear at E5.5–6. EMG and nerve recordings indicate that the neural circuitry involved in producing the basic, coordinated motor pattern of the limbs begins to function early in embryonic devel-

a nipple. The pouchless, gray short-tailed opossum, *Monodelphis domestica*, is born 14–15 days after conception (Kraus and Fadem 1987; Cassidy et al. 1994). McCrady (1938) divided the 13-day prenatal period of the North American opossum into 35 stages. Fertilization begins stage 1, whereas birth, migration to the pouch and attachment to the nipple end stage 35. Stages 2–8 (E2-E3) are cleavage stages, and stages 9–17 (E4-E7) blastocyst stages. The primitive streak appears at stage 16 (E7), at stage 21 (E8) the neural groove has formed, the first somites are formed at stage 22, and the three primary brain vesicles, optic vesicles and otic placodes are present by stage 24 (early E9). At stage 25 (E9), the neural folds start fusion and forelimb plates can be distinguished. Distinct forelimb buds are found at stage 28 (E10). Hindlimb buds cannot be clearly distinguished before stage 30 (E11). The developing animals remain in the pouch for 90 days or more. Of particular interest is the fact that the opossum's hindlimbs are little more than buds at birth and do not become mobile for seven days or more thereafter (Coghill 1938). For convenience, Martin et al. (1978) divided the development of hindlimb motility in *D. virginiana* into three stages: (a) stage I from birth until the first appearance of hindlimb movements (after a week or more in pouch); (b) stage II which begins with the onset of spontaneous hindlimb movements and ends when they can be obviously altered by thoracic transection; and (c) stage III which begins when hindlimb movements can first be affected by thoracic transection (50–60 mm snout-rump length, estimated to be 30–40 days in pouch).

In *Monodelphis domestica*, the forelimbs of the newborn are well developed, with the shoulder, elbow and wrist joints visible, and the five digits individualized and terminated by fine claws. The hindlimbs look like embryonic buds, none of the three joints can be seen, and the toes have not yet individualized. The hindlimbs differentiate according to a proximodistal gradient: at P1–2, the hip joint becomes visible, at P4–5 the knee joint, at P9 the ankle joint, whereas the toes separate from each other at P13. The first autonomous movements of the hindlimbs were observed at P9 (Cassidy et al. 1994). Myelination of fibers innervating the limbs occurs postnatally over a protracted period (Leblond and Cabana 1997). It is possible to study the entire development of hindlimb motility in relation to the development of the CNS postnatally. The locomotor behavior of *Monodelphis domestica* develops also mainly postnatally and can be separated into four broad periods of about 10 days each (Pflieger et al. 1996). During the first period (P1–P10), when the opossum is attached to the mother's nipple, only the forelimbs are capable of rhythmic movements and the hindlimbs are immobile. During the second period of locomotor development (P11–20) the hindlimbs develop rapidly and start moving. In the third period of the opossum's motor development (P21–P30), when the young begins to detach from the mother's nipple, the hindlimbs start to support body weight, and quadrupedal locomotion is possible. In the fourth period (P31-P40), weight support is good, the limbs of both girdles are well coordinated during locomotion, and the opossum's locomotion is more or less adult-like (Pflieger et al. 1996).

4.4
Systems for Staging Rodent and Primate Embryos

For rodents, a system of staging embryos can be used comparable to the Carnegie system for staging primate embryos. This system is derived from Streeter's (1951) approach of staging human embryos into 23 developmental horizons based on their external form and internal characteristics. Each horizon was 2 days apart and numbered XI–XXIII. The transition from embryo to fetus was arbitrarily defined as the point where bone marrow invaded the cartilaginous precursor in the humerus. This event occurs at the end of horizon XXIII. The earlier stages I–X were defined subsequently (Heuser and Corner 1957; O'Rahilly 1973). O'Rahilly and Müller (1987) modified Streeter's system and introduced the term stage instead of horizon as well as Arabic numerals. The Carnegie system has been applied to other primates (Hendrickx and Sawyer 1975; Gribnau and Geijsberts 1981, 1985), and can also be applied to rodent embryos (Theiler 1972; ten Donkelaar et al. 1979). Table 2 summarizes aspects of prenatal development in rodents. The embryonic development of rodents can be arranged up to stage 17 into rather closely corresponding developmental stages as in primates using similar features of external form and internal structure. Some processes are faster in rodents, others certainly slower. The closure of the posterior neuropore is definitely retarded in rodents, whereas the development of the pituitary gland is much faster in rodents than in primates. The transition from embryo to fetus is now defined as the time of closure of the secondary palate (Wilson 1973). These studies show that the order of organogenesis, i.e., the sequence in which individual organs are formed, is basically similar in all mammals studied so far (Table 3). In rats, the neural folds and the first somites are formed at E10, and the rostral neuropore closes at E11. Definite forelimb buds appear at E11–11.5, hindlimb buds at E12. The embryonic period ends at E17. In cats, data are much more restricted. Somite formation starts at E13, the neural tube forms at E14, forelimb buds appear at E18, hindlimb buds at E19, and the palate fuses at E32 (Evans and Sack 1973). Birth occurs at 60–63 days.

In man, embryonic development takes 8 weeks (O'Rahilly and Müller 1987, 1999). The neural groove and folds appear at stage 8 (E18), the three primary brain vesicles at stage 9 (E20), and the neural tube begins to form at stage 10 (E22). The rostral neuropore closes at stage 11 (E24), and the caudal neuropore at stage 12 (E26). Upper limb buds appear at stage 12, and lower limb buds at stage 13 (E28). Secondary brain vesicles form and the cerebral hemispheres become visible at stage 14 (E32). The hand plate develops at stage 15 (E33), the foot plate forms at stage 16 (E37), finger rays are distinct at stage 17 (E41), elbows and toes appear at stage 18 (E44). Embryonic development is complete at stage 23 (E56). The embryonic development of the rhesus monkey, *Macaca mulatta*, closely parallels that in man (Gribnau and Geysberts 1981, 1985).

Table 2. Aspects of prenatal development in rodents. (After ten Donkelaar et al. 1979)

Main embryonic events	Rat[a,d]	Chinese hamster[b]	Mouse[c]	Carnegie stages in primates
Implantation	E5.5–6	E5.5–6	E4.5	4
Organogenesis:				
a) Formation of neural folds; first somites	E10	E10	E8	9
b) Neural folds begin to fuse; turning of embryo	E10.5	E10.5–11	E8.5	10
c) Formation and closure of anterior neuropore	E11	E103/4–11	E9	11
d) Formation of posterior neuropore; definite forelimb bud	E11–11.5	E11–11.5	E9.5	12
e) Closure of posterior neuropore; formation forelimb bud; first lens indentation	E12	E11.5–12	E10	13
f) Forelimb bud two parts; deep lens indentation	E12.5	E12–12.5	E10.5	14
g) Hindlimb bud two parts; closure of lens vesicle	E13	E12.5–13	E11	15
h) Distinct handplate	E13.5	E13.5	E11.5	16
i) Earliest signs of fingers; 3–5 rows of vibrissary papillae	E14	E14	E12	17
j) Closure of secondary palate (marks end of embryonic period)	E17	E16	E15	
Time of delivery (in days)	21–22	20–21	19	

[a] Butler and Juurlink 1987.
[b] ten Donkelaar et al. 1979.
[c] Theiler 1972.
[d] Witschi 1972.

Table 3. Prenatal development in several mammalian species. (After ten Donkelaar et al. 1979)

	Time of implantation	Formation of neural folds	Closure of secondary palate	Time of delivery (in days)
Rodents				
Mouse[m]	E4.5–5	E8	E15	19
Rat[a,b]	E5.5–6	E10	E17	21–22
Chinese hamster[k,l]	E5.5–6	E10	E16	20–21
Carnivores				
Cat[c]	?	~E12	E32	60–63
Primates				
Tupaia[g]	E7	E9	E24	44
Rhesus monkey[d,e]	E9	E20–21	E46–50	164–170
Man[f,h,i,j]	E5–6	E19–21	E56–60	260–280

[a] Coleman 1965.
[b] Edwards 1968.
[c] Evans and Sack 1973.
[d] Gribnau and Geijsberts 1981.
[e] Hendrickx and Sawyer 1975.
[f] Jirásek 1971.
[g] Kuhn and Schwaier 1973.
[h] Nishimura et al. 1968.
[i] Olivier and Pineau 1962.
[j] O'Rahilly and Müller 1987.
[k] Pickworth et al. 1968.
[l] ten Donkelaar et al. 1979.
[m] Theiler 1972.

5 Development of Descending Supraspinal Pathways in Amphibians

Prior to a discussion of the development of descending motor pathways in amphibians, some aspects of the development of the CNS of urodeles and anurans will be discussed. Coghill's and Herrick's studies on the early development of the CNS in the tiger salamander, *Ambystoma tigrinum*, were aimed at determining the relation between early differentiation of the brain and the development of behavior. In anurans, particularly the development of the primary nervous system, a functionally complete, yet transient embryonic nervous system, has been extensively studied.

5.1
Development and Regenerative Capacity of the Motor System in Urodeles

Coghill (1913, 1914, 1929) noted several successive stages in the early development of larval salamanders: the "non-motile" stage, the "early flexure" stage (Harrison's stage 33), the "coil" stage (stage 34/35; see Fig. 7), the stage of the "S-reaction" (stage 35/36), and the "early swimming" stage (stage 37). The early tailbud embryo, though non-motile, has functional muscle cells contacted by motor nerves, and the skin is already innervated by sensory neurons. Gradually, the initial spontaneous and uncoordinated movements yield place to an effective swimming action, in which alternate waves of contraction pass down each side of the embryo and are transmitted sufficiently rapidly to throw the trunk into an S-form. At a stage when the embryo first responds to an external stimulus, most cells in the CNS are still in the neuroblast stage; but comparatively few elements have differentiated and constitute the first reflex arcs (Herrick and Coghill 1915). This *primitive reflex mechanism* (Fig. 7) is formed by Rohon-Beard cells, primitive commissural neurons ("floor plate neurons") and primitive motoneurons. Rohon-Beard cells are transient primary afferent neurons in the dorsal part of the spinal cord receiving proprioceptive and exteroceptive information (Roberts and Clarke 1983; Harper and Roberts 1993). In *Triturus vulgaris*, HRP labeling, together with GABA and glycine immunocytochemistry, revealed nine distinct anatomical classes of spinal cord neurons (Harper and Roberts 1993). Eight of the nine cell classes show a marked resemblance to neurons described in zebrafish (Bernhardt et al. 1990) and *Xenopus* embryos (see Sect. 5.2). The population of descending interneurons extends from the hindbrain through the trunk spinal cord.

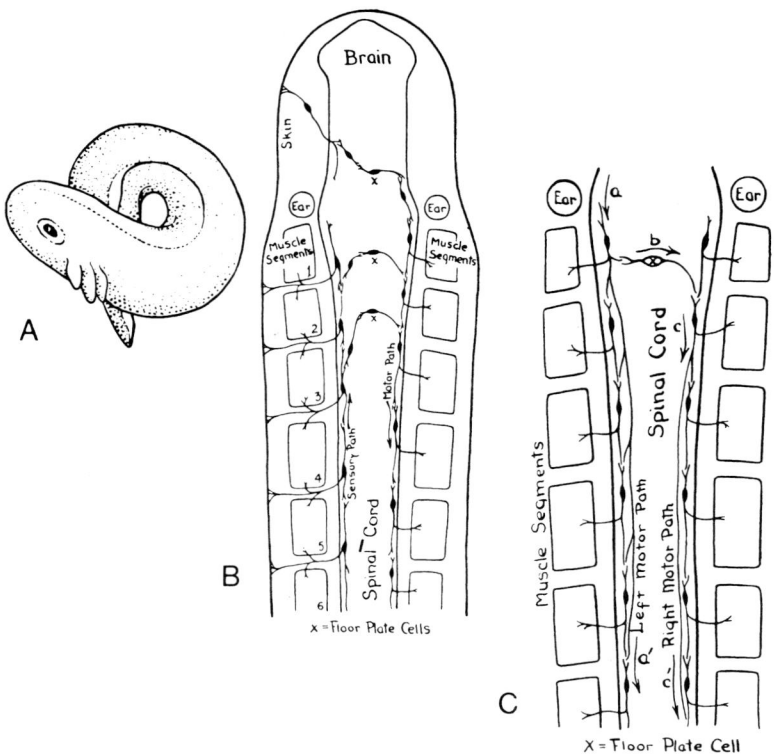

Fig. 7A–C. The primitive reflex mechanism in larval salamanders (after Herrick and Coghill 1915, and Coghill 1929). **A** The coil reaction in *Ambystoma tigrinum* (Harrison's stage 34/35) **B** The presumed mechanism for the coil reaction. A stimulus in front of the ear would excite an afferent neuron which in turn would excite a commissural cell in the floor plate (*x*, floor plate cells). From here the signal is transmitted to the motor tract on the other side **C** The presumed neuromotor mechanism of swimming. The sensory circuitry is omitted. The signal *a–a'* travels down the left motor column and causes the first flexure. Behind the ear, a motoneuron transmits an impulse via the floor plate pathway (*b*), which then excites the second flexure through the right motor column (*c–c'*)

Coghill's (1929) developmental sequence is generally accepted for most anamniotes (Bekoff 1985). In more modern terms, the development of locomotion in urodeles can be divided into: (a) *first movements* (head flexure stage), muscular activity begins with contraction of a few segments of axial musculature in the anterior trunk (or future neck region); (b) the *C-coil stage*, with coordination of ipsilateral myotomes; (c) the *S-wave stage* (bilateral coordination and ipsilateral phase lag); and (d) *swimming*.

In the salamanders *Ambystoma punctatum* and *A. tigrinum*, Coghill (1930, 1931) and Herrick (1937, 1938) described a number of axonal tracts organized in a stereotyped orthogonal pattern (Fig. 8). This *scaffold* is composed of two longitudinal fiber tracts (the dorsolateral and ventrolateral bundles) and three transversely oriented tracts: the telencephalic tract, the dorsoventral diencephalic tract, and the posterior commissure. The *dorsolateral bundle* extends from the isthmus region to the end of

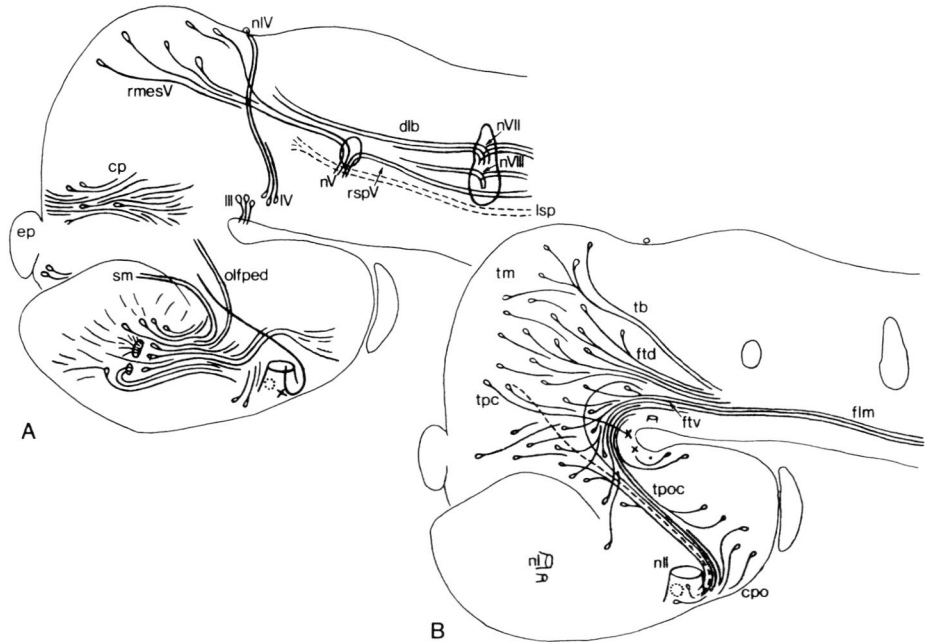

Fig. 8A,B. Herrick's data on early axonal tracts in an early swimming stage *Ambystoma tigrinum* larva (Harrison's stage 37). In **A** olfactory projections, the posterior commissure, and components of the dorsolateral bundle are shown; in **B** the components of the posterior and postoptic commissures, tectobulbar and tectopeduncular tracts, tegmental fascicles and the flm. *cp*, commissura posterior; *cpo*, commissura postoptica; *dlb*, dorsolateral bundle; *ep*, epiphysis; *flm*, fasciculus longitudinalis medialis; *ftd, ftv*, dorsal and ventral tegmental fascicles; *lsp*, lemniscus spinalis; *nI, nII, nIV, nV, nVII, nVIII*, cranial nerves; *olfped*, olfactopeduncular tract; *rmesV, rspV*, mesencephalic and spinal roots of the trigeminal nerve; *sm*, stria medullaris; *tb*, tectobulbar tract; *tm*, tectum mesencephali; *tpc*, crossed tectopeduncular tract; *tpoc*, tract of the postoptic commissure; *III, IV*, motor nuclei of the third and fourth cranial nerves

the spinal cord, and corresponds to Herrick and Coghill's (1915) primitive sensory tract. Initially, the dorsolateral bundle is composed of the axons of the embryonic Rohon-Beard cells. In later developmental stages, axons of spinal ganglion cells and descending fibers of the trigeminal nerve are added to this bundle. Its most rostral part is formed by axons descending from the mesencephalic trigeminal nucleus (Herrick 1938). The *ventrolateral bundle* extends throughout the brain and spinal cord. At spinal levels it is composed at first of the axons of primitive motoneurons forming the primitive motor tract (Herrick and Coghill 1915; Coghill 1929). Descending axons of intersegmental interneurons are added to this tract (Nordlander 1987), and somewhat later this bundle is invaded by axons originating from a group of early-differentiating cells in the synencephalon. The brainstem component of the ventrolateral bundle constitutes the primordial *fasciculus longitudinalis medialis* (flm). The flm is one of the earliest differentiating fiber tracts (Herrick 1937, 1938). The most rostral part of the ventrolateral bundle is formed by the *tract of the postoptic commissure* which is situated in the caudal part of the chiasmal ridge and connects the two sides of the diencephalon. This commissure arises very early in development (Coghill 1930; Herrick 1937, 1938; Wilson et al. 1990).

In the developing urodele spinal cord, longitudinal fiber tract development begins with the growth of neurites into longitudinally oriented spaces near the circumference of the neural tube (Singer et al. 1979). Growing axons are possibly guided along pre-existing substrate routes or *substrate pathways* (Katz and Lasek 1979, 1981; Katz et al. 1980). These observations of cells acting as substrate pathways during development are complemented by a number of demonstrations that CNS axons growing over long distances behave as if they are following discrete pathways in the substrate (see Katz et al. 1980). Their experiments suggested the presence of two stereotyped longitudinal routes: (1) a ventral substrate pathway (the *basal substrate pathway*) which extends into the ventral marginal zone from the caudal diencephalon to the spinal cord, and (2) a dorsal *alar substrate pathway* extending in the alar plate of the developing hindbrain and spinal cord. Transplanted Mauthner cell axons followed the basal substrate pathway (Katz et al. 1980).

In the urodele spinal cord, Singer and co-workers (Egar and Singer 1972; Nordlander and Singer 1978; Singer et al. 1979) have discovered a prototype substrate pathway in the preneural "ependymal tunnels" which guide axons in the regenerating and developing spinal cords. In regenerating newt (*Triturus viridescens*) tails, ependyma-like germinal cells form a new neural tube and radial cell processes of the regenerating ependyma form channels between them (Fig. 9) which are subsequently invaded by growing axons (Egar and Singer 1972; Nordlander and Singer 1978). The germinal neuroepithelium of the amphibian embryo and larva patterns these longitudinal neural tracts in a similar manner (Singer et al. 1979; Nordlander and Singer 1982a,b). In *Xenopus laevis*, Nordlander and Singer (1982a,b) showed that earliest fiber tract development in the tail spinal cord occurs by the ingrowth of axons into small but

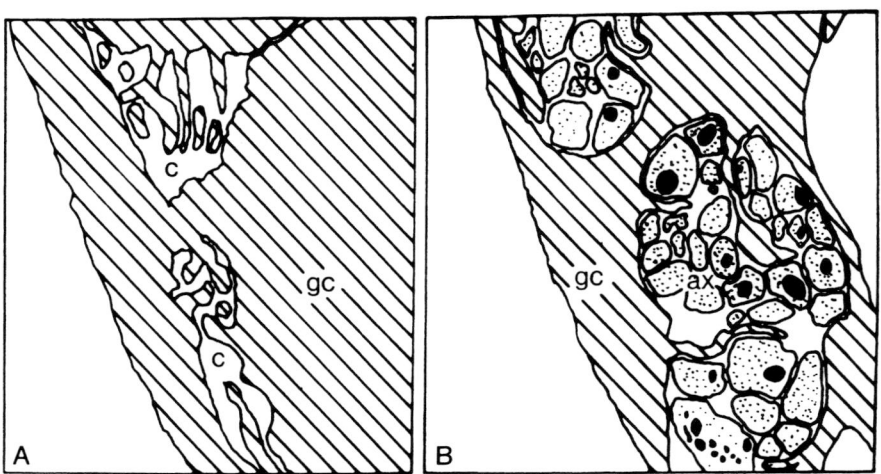

Fig. 9. Drawing of regenerating newt spinal cord showing intercellular spaces (*c*) between radial processes of germinal cells (*gc*) **A** *before* axons (*ax*) have reached them, and **B** *after* axons are growing within them (based on data by Nordlander in Katz et al. 1980)

discrete longitudinally oriented spaces between neighboring cells of the neural tube. These spaces, which are located at the periphery of the cord, are empty at the most primitive stages but with further development they are infiltrated by increasing numbers of neurites.

Salamanders are the only limbed vertebrates capable of recovering locomotor function following transection of all levels of the spinal cord. The best studied species are the axolotl and the newt *Notophthalmus* (*Triturus*) *viridescens*, salamanders capable of regenerating various body regions including limbs (e.g., Goss 1969; Singer and Caston 1972; Tank and Holder 1981; Muneoka et al. 1989) and the spinal cord (Kirsche 1956; Butler and Ward 1967; Egar and Singer 1972; Nordlander and Singer 1978; Simpson 1983; Clarke et al. 1988; Holder and Clarke 1988; Davis et al. 1989a, 1990). The time course of salamander spinal cord regeneration and recovery of locomotion was studied primarily in *N. viridescens* (Davis et al. 1989a, 1990). The extent of the regeneration was much more complete at tail levels than in other regions of the spinal cord. Following ablation of the tail, the regenerate contains a spinal cord complete with sensory ganglia, motoneurons and interneurons, and spinal tracts (Piatt 1955). Ablation or transection of the brachial or thoracic spinal cord results in a regenerated spinal cord that is often only a fraction of the size of the normal spinal cord, contains hardly any neurons, has greatly reduced white matter, and little or no neuropil (Butler and Ward 1965, 1967; Stensaas 1983; Davis et al. 1989a). In *N. viridescens*, ependymal regrowth into the tail regenerate occurs followed by ependymal differentiation including the formation of channels (Fig. 9) which are invaded by descending growing axons (Egar and Singer 1972; Nordlander and Singer 1978). Sensory axons in the dorsal funiculus do not regenerate (Stensaas 1983). The presence of vimentin, normally present in immature glial cells, in radial glia of adult newts (Zamora and Mutin 1988) also fits in with the view that the ependyma of the adult newt expresses potentialities for development similar to those exhibited by the germinal neuroepithelium of the embryo (Singer et al. 1979). In urodeles, many of the glial cells express keratin intermediate filaments, and a correlation exists between the regenerative capacity of axons and the presence of keratin filaments in neighboring glia (Holder et al. 1990; O'Hara et al. 1992). Moreover, CNS myelin of the axolotl is a growth-permissive substrate in vitro (Lang et al. 1995). It does not exhibit neurite growth inhibitors.

The spinal circuitry in the regenerated spinal cord was studied in *Notophthalmus viridescens* (Davis et al. 1989a, 1990). The animals received a complete transection at the thoracolumbar junction, abolishing all spontaneous coordinated hindlimb and tail movements. Animals exhibited walking and swimming within 60 days at which time a plug of HRP was inserted into a gap in the spinal cord made by a transection some 10 mm caudal to the first lesion. On average the number of HRP-labeled brainstem neurons in regenerated animals was about 40% of that found in normal animals. The number of labeled propriospinal neurons in the brachial spinal cord was within the range of normal animals. Regenerated salamanders had HRP-labeled cells in all regions of the brain stem including the red nucleus, the interstitial nucleus of the flm and the reticular formation. Thus, in salamanders, initially projecting descending supraspinal and propriospinal axons have the ability to regenerate at least 10 mm past a complete transection (Davis et al. 1989a). Recovery of coordinated swimming was only observed in salamanders in which descending supraspinal and (long) propriospinal axons were present at the level of the lumbar enlargement (Davis et al. 1990).

These data indicate that recovery of locomotion is dependent on the re-establishment of descending input.

Holder and Clarke (1988) suggested that functional recovery in the CNS is linked directly to the developmental stage of the region or the system that is lesioned. In later studies (Holder et al. 1991) they presented evidence for continuous neurogenesis and growth in the motor system of the axolotl. Repair in this system may be linked to the maintenance of cues within the environment of regenerating axons which guide them to their appropriate targets. This view suggests that the ability to regenerate is an inevitable consequence of continuous growth.

5.2
Development of Anurans: Two Motor Systems

The CNS develops from the neural plate, which in *Xenopus laevis* is a bilayer of ectodermal cells in the middorsal region (Schroeder 1970; Hartenstein 1989). The neural plate invaginates to form the neural tube, forming a pseudostratified neuroepithelium (Hartenstein 1989). Subsequently, flexures and constrictions subdivide the neural tube, so that by stage 25 the prosencephalon with outgrowing eye vesicles, the mesencephalon, the rhombencephalon and the spinal cord can be distinguished. During the larval development of *X. laevis* the CNS develops with a caudorostral gradient, i.e., the spinal cord and the rhombencephalon develop first, followed by the mesencephalon, the diencephalon and the telencephalon, which appear to have their developmental climax during metamorphosis (Nieuwkoop and Faber 1967).

At hatching, the anuran CNS comprises two classes of cells: (1) neurons and glial cells which have become postmitotic at some point during early embryogenesis and form a differentiated circuit which controls early larval behavior (Roberts and Clarke 1982; Roberts 1989, 1990; Hartenstein 1989, 1993); (2) the still undifferentiated, proliferating precursors which over the larval period will produce the majority of cells found in the adult CNS. For the early differentiating population the term "primary neurons" will be used, and "secondary neurons" for the later appearing neurons. Primary sensory neurons (Rohon-Beard cells, extramedullary neurons) and primary motoneurons differ from their secondary counterparts not only in birthday (Lamborghini 1980; van Mier 1986), but also in size, position and axonal arborization (Forehand and Farel 1982b; Roberts and Clarke 1982; van Mier et al. 1985; Nordlander 1986; ten Donkelaar and de Boer-van Huizen 1991). Between stages 10 and 21 as much as 75% of the primary motoneurons are generated. In this period already some 30% of the secondary motoneurons innervating the limb muscles are born and persist through metamorphosis (van Mier 1986).

Acetylcholinesterase seems to be a good marker for identifying primary neurons at early stages of differentiation (Moody and Stein 1988). Lineage tracing (Hirose and Jacobson 1979; Jacobson and Hirose 1981) and pulse labeling experiments (Lamborghini 1980; Hartenstein 1989) showed that after gastrulation there is a wave of mitosis. Most cells of the neural plate undergo a single division during this wave. After this first division, many cells leave the cell cycle and differentiate as primary neurons (Hartenstein 1989). In stage 35/36 hatching larvae, clones of primary neurons usually contain only two cells. The remainder, most of which arise from the superficial layer of the neural plate, are predominantly the precursors of secondary neurons. They are mitotically quiescent until stage 20, and then undergo another one to two rounds of division during embryonic life. Secondary precursors and primary neurons are never of the same clone. By the neural plate stage, separate precursors seem to exist for primary and secondary neurons (Hartenstein 1989). Jacobson and Huang (1985) injected HRP into blastomeres at the 32-cell stage. The tracer was transmitted during mitosis to all the descendants and could be seen up to a week later in well-differentiated cells, including neurons and their peripheral targets. The tracer entered the outgrowing neurites and showed the position and directions of initial outgrowth of axons as well as dendrites, the pattern of neurite branching, and the relation between axon terminals and specific targets. All types of nerve fibers studied grew by the most direct pathway, apparently without errors of initial outgrowth, pathway selection, or target selection. Wholemount antibody labeling techniques and HRP backfilling in the embryonic *Xenopus* CNS between stages 22 and 35/36 showed that in the spinal cord the first neurons to differentiate are the Rohon-Beard cells, followed by ventral neurons with descending axons (descending interneurons and motoneurons) and lateral interneurons with commissural axons (Hartenstein 1993). The cell bodies and axons of these primary cell populations form dorsal, ventral, and lateral columns, respectively. Both the ventral and lateral columns continue uninterruptedly in the brain stem.

Cells in the CNS appear to go through several phases during their differentiation. As shown by HRP labeling (Jacobson and Huang 1985) and transmitter immunohistochemistry (van Mier et al. 1986; Gallagher and Moody 1987; Roberts et al. 1987, 1988), from the moment cell division stops, an axon is formed followed by dendrites which emerge from the cell body. Usually these dendrites then lose their embryonic features, and become numerous and thinner. At the beginning of the differentiation phase the production of the cell-specific neuroactive substances takes place. Initial outgrowth is in a specific direction for each class of neuron (Roberts 1988). Within the marginal zones of the spinal cord and hindbrain axon outgrowth is limited as regards orientation to *ascending longitudinal* (e.g., ascending interneurons), *descending longitudinal* (e.g., raphespinal neurons), and *ventral circumferential* (commissural interneurons). *Each* neuron is a pioneer and has initial outgrowth in an appropriate direction *without* contact with its peers (Roberts 1988). In the early CNS of *Xenopus laevis* there seems to be little evidence that the first cells to differentiate within a class are different from those which appear later. In the tail spinal cord, each axon of sensory ganglion cells is able to find its way *independently* (Nordlander et al. 1991).

The HNK-1 antigen, a carbohydrate moiety bound to many cell adhesion and recognition molecules, is implicated in cell-cell and cell-substrate interactions during neural development. It marks earliest axonal outgrowth in *Xenopus* (Nordlander 1989, 1993). Neurons of the earliest circuits begin to express the antigen shortly after they are born and begin to differentiate. A neuron-specific β-tubulin marks growing axons during the embryonic period (Hartenstein 1993; Moody et al. 1996). At stages 21–28, the pioneering axons of Rohon-Beard, commissural, primary motor, and trigeminal ganglion cells were distinctly stained in the axonal scaffolds that they formed in the embryonic brain. Throughout swimming and premetamorphic stages, neuronal cells in all brain regions became immunoreactive as they differentiated and extended axons. Neurons expressed detectable levels of this class II β-tubulin protein only beginning at the onset of overt axon outgrowth.

The *"primary motor system,"* responsible for the reactions of a tadpole to stimuli, comprises only a limited number of distinct classes of neurons (Roberts and Clarke 1982; Roberts 1989, 1990; Hartenstein 1993) including Rohon-Beard cells, primary motoneurons and several classes of spinal interneurons (Fig. 10). Immunocytochemistry and tract-tracing studies greatly helped in classifying these tadpole interneurons (Roberts and Clarke 1982; Dale et al. 1986; Roberts et al. 1987, 1988; Roberts and Sillar 1990; Heathcote and Chen 1993, 1994). The extensive electrophysiological studies by Roberts and co-workers, focussed on hatchling (stage 37/38) *Xenopus laevis* tadpoles, unraveled much of the mechanisms underlying the flexure and swimming responses (Roberts 1990; Roberts et al. 1997). Swimming and "fictive" swimming can be initiated by touching the skin. This excites sensory Rohon-Beard cells (Clarke et al. 1984), which in turn excite spinal sensory interneurons including dorsolateral commissural interneurons (Clarke and Roberts 1984; Sillar and Roberts 1988; Roberts and Sillar 1990). The decussating axons of these interneurons carry excitation to the opposite side of the spinal cord (Sillar and Roberts 1988). This pathway produces a crossed flexion reflex by exciting motoneurons contralateral to the stimulus (Clarke and Roberts 1984; Sillar and Roberts 1988; Roberts and Sillar 1990). Intracellular recordings, dye injections and pharmacological studies in *X. laevis* have shown that: (1) "commissural" interneurons are active during swimming and produce reciprocal glycinergic inhibition of rhythmically active neurons on the opposite side of the cord including other "commissural" interneurons (Soffe et al. 1984; Dale 1985; Soffe 1987); (2) excitatory interneurons, almost certainly "descending" interneurons, are active during swimming and produce long-lasting "glutaminergic" excitation of all rhythmically active neurons on the same side of the cord (Dale and Roberts 1984, 1985; Roberts and Alford 1986); and (3) motoneurons show the same pattern of activity as the rhythmic interneurons (Roberts and Kahn 1982; Soffe and Roberts 1982a,b; Perrins and Roberts 1995b,c).

Roberts' data (see Fig. 11) suggest the presence of a rhythm-generating network (*central pattern generator* or CPG) in the embryonic spinal cord, very similar to Brown's (1911) half-center hypothesis. Excitatory interneurons that fire during swimming excite each other as well as the motoneurons and inhibitory interneurons that form the remaining components of the swimming network (Roberts 1990; Fetcho 1992). Motoneurons are not pure output devices, however. They make two types of synapses with other motoneurons (Perrin and Roberts 1995a–c): the first are *chemical*, cholinergic synapses activating postsynaptic nicotine receptors, the second are *electrotonic* synapses. Both types of synapse appear to be made from the longitudinal central axons of the motoneurons (Roberts and Walford 1996). Motoneurons may be an integral part of the *Xenopus* CPG for swimming (Perrin and Roberts 1995b,c; Roberts et al. 1997). Reticulospinal neurons participate in early swimming (Roberts and Alford 1986; van Mier and ten Donkelaar 1989). *Sensory gating* mechanisms are used to ensure that adaptive and appropriate motor responses to apparent input occurs during all phases of the movement (Sillar 1991). Premotor interneurons, rhythmically active during locomotion, as well as 'sensory' interneurons appear to be intimately involved in sensory gating, receiving synaptic inputs from the spinal rhythm generator to gate the flow of sensory information in the spinal cord (Sillar and Roberts 1992). Swimming can be stopped by pressure on the cement gland on the head (Boothby and Roberts 1992). Trigeminal sensory neurons then excite inhibitory, GABAergic reticulospinal neurons which turn off swimming.

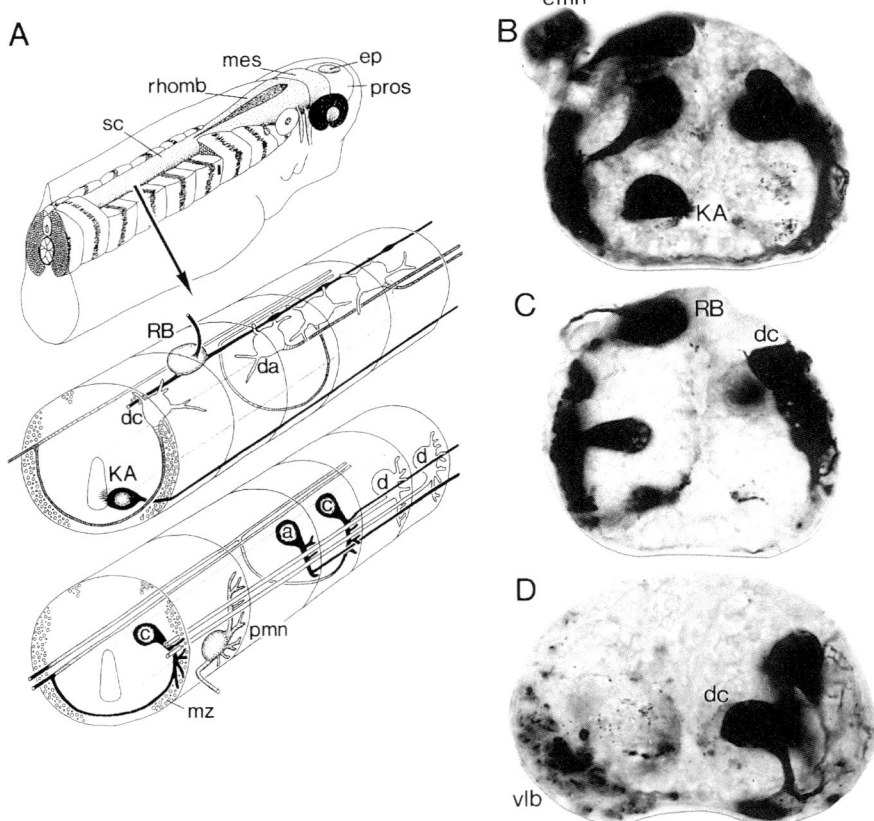

Fig. 10. A Organization of the spinal cord in a stage 37/38 *Xenopus laevis* embryo (after Roberts 1990) with the neurons of the sensory pathways and the motor system. Dark neurons are inhibitory neurons **B–D** Photomicrographs showing some of its cell populations (from ten Donkelaar and de Boer-van Huizen 1991; **B,C**, stage 35; **D**, stage 41). *a*, ascending interneuron; *c*, commissural interneuron; *d*, descending interneuron; *da*, *dc*, dorsolateral ascending and commissural interneurons; *emn*, extramedullary neuron; *ep*, epiphysis; *KA*, Kolmer-Agduhr neuron; *mes*, mesencephalon; *mz*, marginal zone; *pmn*, primary motoneuron; *pros*, prosencephalon; *RB*, Rohon-Beard cell; *rhomb*, rhombencepalon; *sc*, spinal cord; *vlb*, ventrolateral bundle of ascending and descending fibers

In *Xenopus laevis*, spinal ganglia can be observed by stage 39 (Hughes and Tschumi 1958; Nieuwkoop and Faber 1967) as aggregates of 6–10 neuroblasts. By stage 44, the number of cells in each ganglion had increased threefold, and dorsal roots become distinguishable. Dorsal roots are well developed by stage 47. Around stage 45, the first neurites emerge from dorsal root ganglion (DRG) cells related to the hindlimbs, and start to grow towards the hindlimb buds and the spinal cord (van Mier and ten Donkelaar 1988). At stage 48, the first primary afferent fibers enter the medial part of the dorsal funiculus of the lumbar cord. Data in the bullfrog, *Rana catesbeiana*, indicate that in anurans the development of thoracic DRG cells and their projections precedes that of brachial and lumbar ganglia (Smith and Frank 1988a,b). In *X. laevis*,

Fig. 11. Rhythm generation in *X. laevis tadpoles* as demonstrated by Roberts and co-workers (based on Arshavsky et al. 1993, and Roberts et al. 1997). A dorsal view of the brain and rostral spinal cord (the first six postoptic myotomes) of a stage 37/38 larva, and an nlarged part showing the labeled neurons after HRP application at the left half of the fourth spinal segment are depicted. *Dark cells* indicate the descending interneurons in the rhombencephalic reticular formation. These cells drive the half centers present in the spinal cord. Motoneurons are not pure output devices as previously thought, but make two types of synapses with other motoneurons: (1) chemical, cholinergic synapses, and (2) electrotonic synapses (*resistor sign*). Abbreviations: *CIN*, commissural (inhibitory) interneurons; *DINs*, descending interneurons; *EIN*, excitatory interneurons; *mes*, mesencephalon; *Mn*, motoneurons; *oc*, otic capsule

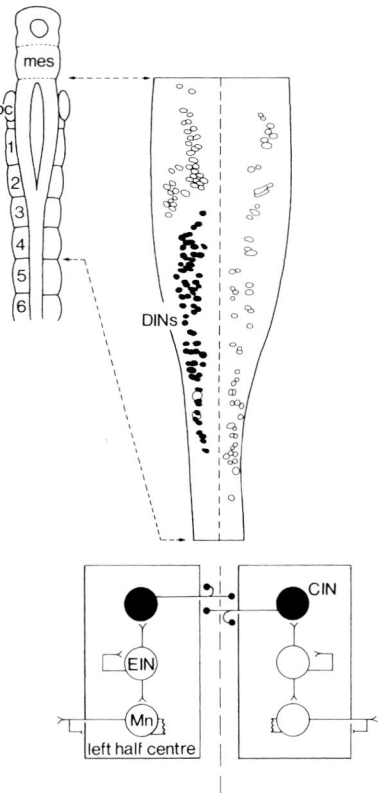

around stage 48 DRG cells could be labeled from the developing dorsal column nucleus, but only from non-limbbud-innervating DRG (ten Donkelaar and de Boer-van Huizen 1991). Gradually, also the limbbud-innervating ganglia give rise to ascending collaterals, so that by stage 53 all spinal ganglia send ascending collaterals to the brain stem (for data on ranid frogs see Forehand and Farel 1982c). The first mesodiencephalic projections from the dorsal column nucleus were found at stage 51 (Muñoz et al. 1993), i.e., not much later than projections of DRG cells reach the dorsal column nucleus. At stage 50, spinocerebellar projections appear that arise from cervical and lower thoracic/upper lumbar levels (van der Linden and ten Donkelaar 1987; van der Linden et al. 1988).

The development of the relationship between dorsal root fibers and motoneurons has been extensively studied in the larval bullfrog spinal cord (Liuzzi et al. 1985; Smith and Frank 1988a, b). In the tail of *Xenopus laevis*, the DRG are rather diffuse, and their constituent cells are divided into two groups, one close to the cord within its pigmented meningeal sheath, and the other ventrolateral to the first, between the notochord and the inner surface of the myotome (Hughes and Tschumi 1958). Nordlander et al. (1988) showed that dorsal roots are *absent* from the tails of *X. laevis* larvae. Sensory afferent fibers instead enter the spinal cord via the ventral roots. The caudal (tail) spinal cord in *Rana* and *Xenopus* tadpoles differs markedly (Nishikawa and Wassersug 1988).

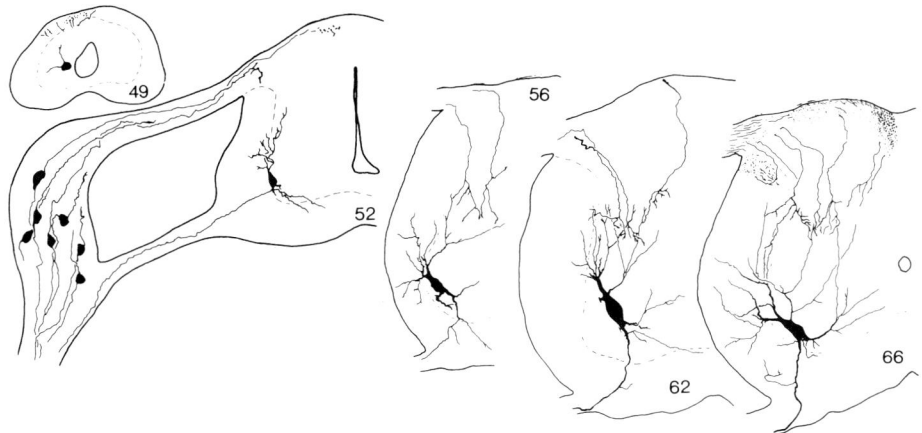

Fig. 12. A developmental sequence of the primary afferent fibers and the lateral motoneurons in the lumbar spinal cord observed after HRP application to the hindlimb buds of *Xenopus laevis* larvae ranging from stage 49 to 66 (after van Mier and ten Donkelaar 1988)

At stage 48, shortly after the hindlimb bud arises (stage 46, early metamorphosis), the first motoneurons related to this developing extremity could be labeled in the ventrolateral part of the lumbar spinal cord (van Mier et al. 1985). At first these *secondary motoneurons* bear only a few dorsal dendrites of which only the tips reach out into the adjacent white matter (Fig. 12). Already at stage 50, these dorsal dendrites have invaded the whole dorsolateral part of the marginal zone. Also the first ventral dendrites were observed at this stage. Later, at stage 53/54 some ventral and lateral dendrites have also reached the white matter. At these early metamorphic stages already some primary afferent fibers were found to make contact with the dorsomedial dendrites (van Mier and ten Donkelaar 1988). These data suggest a "waiting period" between the arrival of dorsal root fibers in the spinal white matter and the extension of collateral branches into the spinal gray matter as found by Smith and Frank (1988a) in *Rana catesbeiana*. At stage 58, for the first time recurrent axon collaterals were found extending into the ventromedial part of the marginal zone. The development of hindlimb-innervating motoneurons seems to be characterized by two phases: (1) the establishment of contacts between motoneurons and their target muscles, and (2) the subsequent formation of connections of these motoneurons with interneurons, dorsal root fibers and descending supraspinal fibers.

5.3
Neurogenesis of Brainstem Motor Nuclei:
Birthday Data in Xenopus laevis

Many of the large basal plate neurons in the brainstem reticular formation are generated during early gastrulation (Lamborghini 1980; Hartenstein 1989). The Mauthner cells are also generated during this period (Vargas-Lizardi and Lyser 1974). Vestibular neurons are also born very early in development, but red nucleus neurons undergo

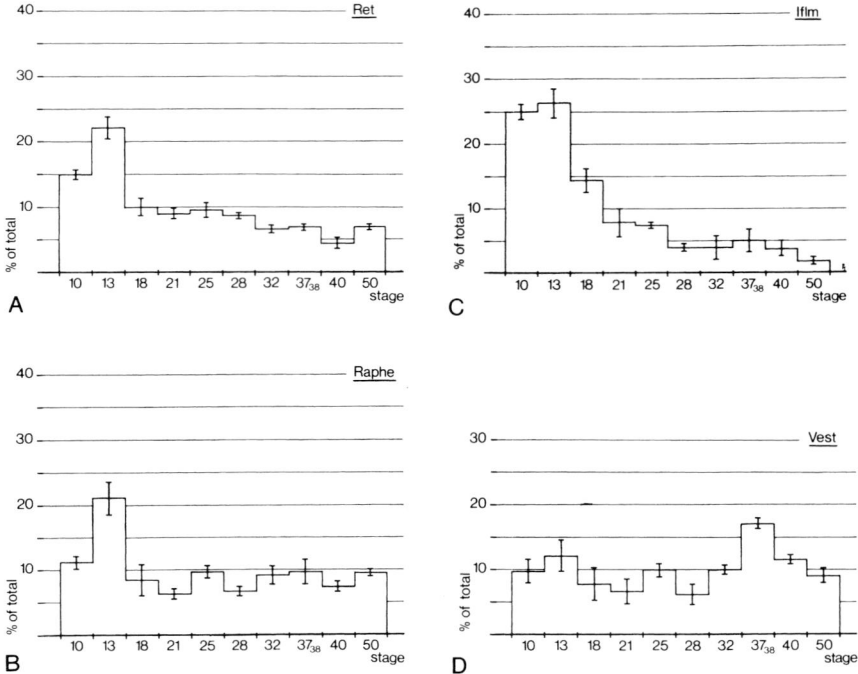

Fig. 13. The number of heavily labeled cells after [³H]-thymidine injections in embryos and larvae of *Xenopus laevis* at various stages of development in **A** the reticular formation, **B** the raphe nuclei, **C** the interstitial nucleus of the flm, and **D** the vestibular nuclear complex (from van Mier 1986)

their final mitosis much later (van Mier 1986). With [³H]-thymidine autoradiography van Mier (1986) studied the "birthday" of reticular, vestibular and rubral neurons in the brain stem. The combined use of [³H]-thymidine autoradiography and HRP histochemistry was used to correlate the time of origin of neurons in the reticular formation with the appearance of their spinal projections.

Neurogenesis in the brain stem starts very early. In embryos which received a [³H]-thymidine injection at stage 10 or 13, and were allowed to complete development until stage 66, many heavily labeled neurons were found in the brain stem. In the reticular formation, at least 35% of all the neurons that were born between stages 10 and 50 originate between stages 10 and 13 with a peak production of 22% at stage 13 (Fig. 13A). After this a gradual decrease can be observed in the number of reticular neurons born to as low as 4% at stage 40. At stage 50 a small but significant increase was observed in the cell production in the rostral part of the reticular formation. No clear caudorostral gradient was found. Slightly more large than small reticular neurons arise during the early gastrula stages (stages 10–15), whereas at later stages (18–50) many small reticular neurons are born. Neurons in the nucleus raphes are also born during early gastrulation with a peak value of 22% at stage 13 (Fig. 13B). At least 50% of the cells of the interstitial nucleus of the fasciculus longitudinalis medialis (flm) born between stages 10 and 50 are generated during stages 10 and 13 (Fig. 13C). In the production of vestibular neurons a maximum of 17% was observed

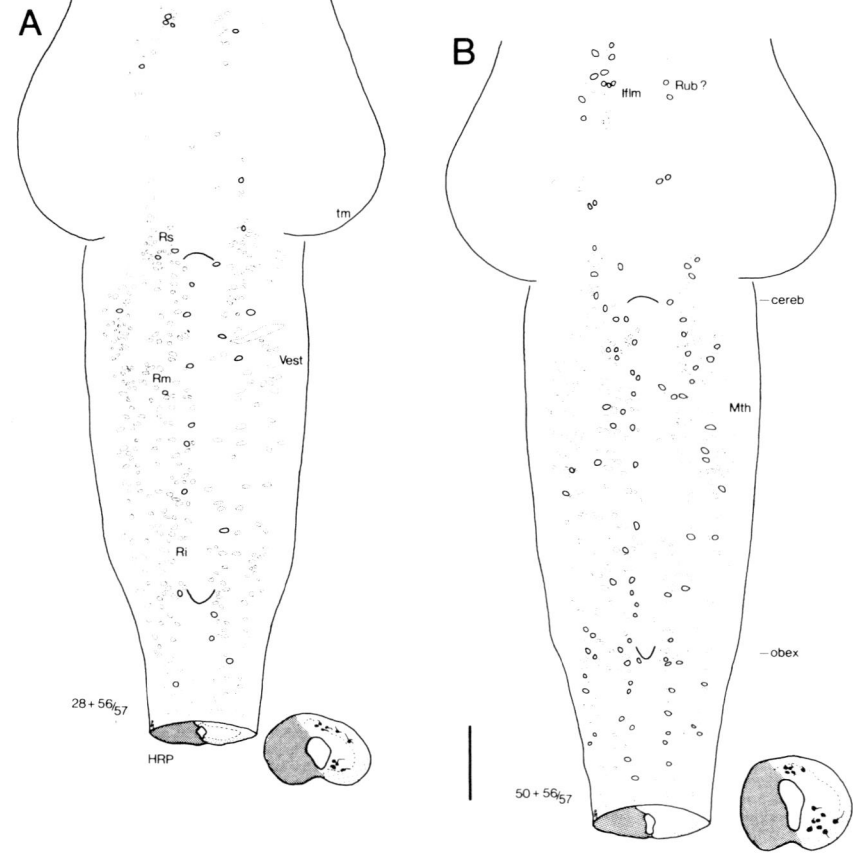

Fig. 14. The distribution of HRP labeled cells (indicated by *heavy, uninterrupted lines*) in the brain stem of *Xenopus laevis* after a [³H]-thymidine injection at stage 28 **A** and 50 **B**, respectively, and HRP application at stage 56/57 in both experiments. The cells indicated by *interrupted lines* are only backfilled with HRP (from van Mier 1986). *cereb*, cerebellum; *Iflm*, interstitial nucleus of flm; *Mth*, Mauthner cell; *Ri, Rm, Rs*, inferior, middle and superior reticular nuclei; *tm*, tectum mesencephali; *Vest*, vestibular nuclei

at stage 37/38 (Fig. 13D). Before and after this maximum the generation of vestibular neurons appeared to be more or less constant between 6% and 10%. Between stages 10 and 18 many neurons were born in the rostral two-third of the vestibular nuclear complex. Between stages 18 and 37/38 neurons are born throughout the entire vestibular nuclear complex. At stages 37/38 and 40 a slight increase in cell production could be observed in the caudal part of the vestibular nuclear complex.

In larvae which received a [³H]-thymidine injection at stage 28 and a spinal HRP application at stage 56/57, a small number of double labeled cells were found more or less evenly distributed throughout the brainstem reticular formation including the interstitial nucleus of the flm (Fig. 14A). Only a few double labeled cells were found in the vestibular nuclear complex. Approximately 7% of the total number of HRP labeled cells was heavily labeled with silver grains. These double labeled cells are born at or close to stage 28 and project to the spinal cord at stage 56/57. All other retrogradely labeled neurons presumably are born at other stages in development. After application of HRP to the spinal cord of stage 56/57 larvae (Fig. 14B) which received a [^{3}H]-thymidine injection at stage 50, as much as 26% of the HRP labeled cells was also heavily labeled with silver grains. These double labeled cells are probably born at or close to stage 50. At this stage of development, the double labeled cells were not restricted to the brainstem reticular formation, but were also observed in the vestibular nuclear complex and in the nucleus ruber.

5.4
Tract-Tracing Data in Anurans

5.4.1
Retrograde Tracer Data

Of the classes of propriospinal neurons in *Xenopus laevis* discussed in Sect. 5.2, one extends as a column into the reticular formation of the hindbrain (Roberts and Alford 1986), where they would be called *reticulospinal neurons* (for data in *Triturus vulgaris* embryos see Harper and Roberts 1993). Roberts and co-workers (Roberts et al. 1987; Roberts 1988, 1989) showed that many "mid-hindbrain" reticulospinal neurons are GABAergic cells. Excitation of this cell population inhibits spinal rhythm generation and stops swimming (Boothby and Roberts 1992). They can be stained as early as stage 25 and develop as a compact group with descending ipsilateral and contralateral axons. HRP studies (van Mier and ten Donkelaar 1984; Nordlander et al. 1985; Hartenstein 1993) showed that in *X. laevis*, already prior to the "swimming" stage, reticulospinal fibers arising in the medulla oblongata innervate the spinal cord. The first descending supraspinal fibers penetrate the spinal cord at about stage 28 (Kevetter and Lasek 1982; van Mier and ten Donkelaar 1984; Fig. 15). Hartenstein (1993) noted the presence of eight rhombomeres in the hindbrain. The relationship between these rhombomeres and cranial nerves appears to be similar to that shown for other vertebrates (Keynes and Lumsden 1990). Neurons in the hindbrain labeled by HRP backfilling from the spinal cord were not grouped into metamerically repeated clusters (Roberts and Clarke 1982; van Mier and ten Donkelaar 1984; Nordlander et al. 1985; Hartenstein 1993) like in the zebrafish (Kimmel et al. 1982, 1985; Metcalfe et al. 1986; Mendelson 1986a, b) and the goldfish (Lee et al. 1993). Instead, they formed more or less continuous lateral and ventral columns. Neurons of both columns were labeled with an anti-acetylated tubulin antibody, and could be reliably charted at early developmental stages (Hartenstein 1993). The ventral column is formed by neurons in the interstitial nucleus of the flm and the hindbrain reticular formation, whereas the column of GABAergic cells described by Roberts and co-workers forms the lateral column. At no stage any overt segmental, i.e., rhombomeric, clustering of reticulospinal neurons is apparent. The axons of the interstitial nucleus of the flm and hindbrain reticulospinal axons pioneer the early fasciculus longitudinalis medialis.

In HRP experiments (van Mier and ten Donkelaar 1984) the first brainstem neurons were found in the nucleus reticularis inferior and medius at stage 28, when the

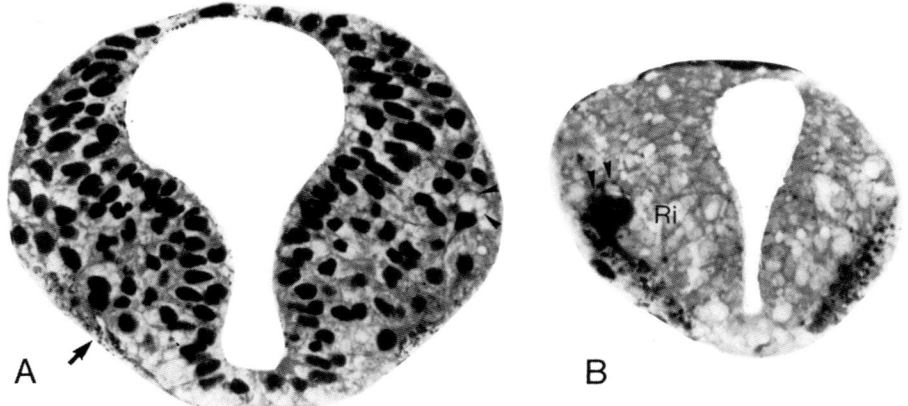

Fig. 15A,B. Photomicrographs showing the ingrowth of descending supraspinal fibers into the spinal cord of a stage 28 *Xenopus laevis* embryo. **A** A Rager-stained tranverse section of the brain stem near the obex region showing fibers in the marginal zone (*large arrow*), ×550 **B** HRP-labeled reticulospinal neuron with vacuoles (*arrowheads*) in the nucleus reticularis inferior (*Ri*), and labeled fibers in the caudal brain stem, ×450. (From van Mier and ten Donkelaar 1984)

neural tube has just closed (Fig. 16A). At this stage of development, the first, uncoordinated swimming movements can be observed. At stage 30/31 (Fig. 16B), both Mauthner cells project to the spinal cord as well as the interstitial nucleus of the flm. Towards stage 35/36 (Fig. 16C), a more extensive reticulospinal innervation of the spinal cord appears, now including cells in the nucleus reticularis superior. These projections all arise before the limb buds appear, and reach the tail spinal cord by stage 37 (Nordlander et al. 1985). Supraspinal axons arrive at the tail spinal cord after longitudinal axons of local spinal neurons are already present (Nordlander 1984). The presence of well-established longitudinal pathways in the tail spinal cord prior to the ingrowth of supraspinal axons suggests that these pathways may serve as the substrate and possibly offer guidance for the caudal growth of supraspinal axons (Nordlander et al. 1985). At stage 43, the pattern of reticulospinal projections appears to be complete with the presence of labeled neurons in the nucleus reticularis isthmi (Fig. 16D). It is unclear whether this cell population includes a locus coeruleus, which in adult *Xenopus laevis* is found dorsolateral to the nucleus reticularis isthmi (ten Donkelaar et al. 1981; González et al. 1993; Marín et al. 1996). From stage 43 on, the number of HRP-positive neurons is steadily increasing. At stage 47/48 (Fig. 16E), when the hindlimb buds appear, the descending projections to the spinal cord are comparable with the adult situation except for the absence of rubrospinal and hypothalamospinal projections.

In limbbud stages, HRP slow-release gels were implanted into the spinal cord (ten Donkelaar and de Boer-van Huizen 1982). At stage 50 (Fig. 17A), apart from the extensive reticulospinal projections, two populations of vestibulospinal neurons were found: the rostral part of the ventral octaval nucleus projects ipsilaterally, and its caudal part contralaterally. Just above the nucleus reticularis isthmi, the coeruleospinal projection arises. This monoaminergic cell group is much better developed in later

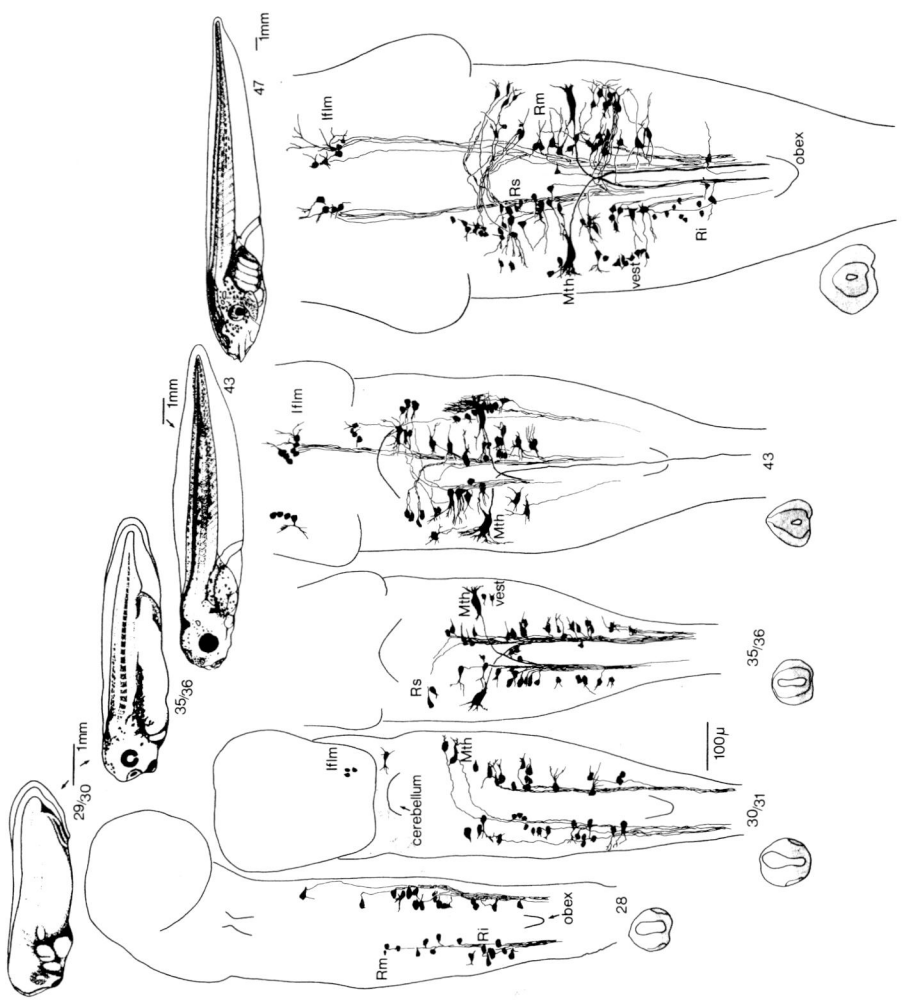

Fig. 16. The distribution of retrogradely labeled neurons in the brain stem at various stages of development in *Xenopus laevis* as found in wholemount preparations (ventral views). The transverse sections show the HRP distribution at the site of application. *Iflm*, interstitial nucleus of flm; *Mth*, Mauthner cell; *Ri, Rm, Rs*, nucleus reticularis inferior, medius, and superior, respectively; *vest*, vestibular nuclear complex. (From van Mier and ten Donkelaar 1984)

stages, e.g., in stage 55 (Fig. 17B). In this series of experiments, rubrospinal fibers were not found to reach the lumbar cord before stage 58, i.e., when the hindlimbs are used for locomotion (ten Donkelaar and de Boer-van Huizen 1982). Hypothalamospinal fibers also innervate the spinal cord relatively late in development, i.e., not before stage 57 (ten Donkelaar and de Boer-van Huizen 1982). Rubrospinal tract neurons start invading the spinal cord at stage 48 (ten Donkelaar et al. 1991). Retrograde tracer (HRP, fluorescent dextran amines: FDA, RDA) data suggest that (1) part of the su-

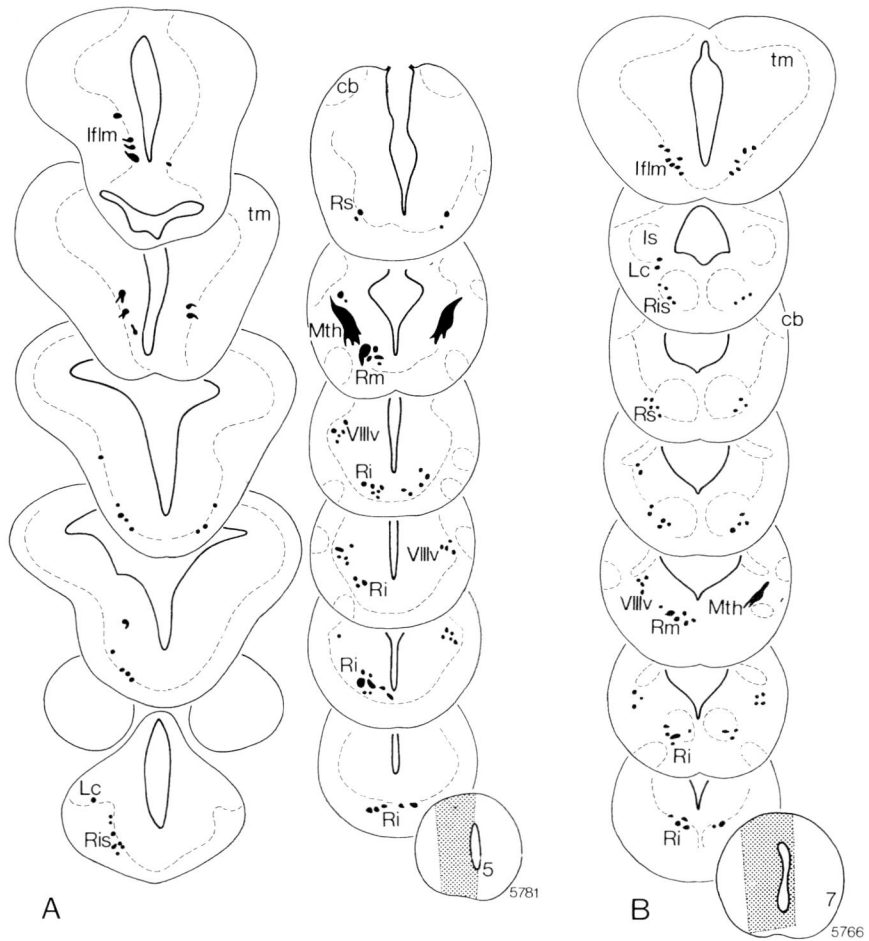

Fig. 17A,B. The distribution of retrogradely labeled cells in the brain stem of limbbud stages of the clawed toad, *Xenopus laevis*. **A** stage 50 **B** stage 55. *cb*, cerebellum; *Iflm*, interstitial nucleus of flm; *Is*, nucleus isthmi; *Lc*, locus coeruleus; *Mth*, Mauthner cell; *Ri*, *Ris*, *Rm*, *Rs*, nucleus reticularis inferior, isthmi, medius, and superior, respectively; *tm*, tectum mesencephali; *VIIIv*, ventral nucleus of VIIIth nerve (After ten Donkelaar and de Boer-van Huizen 1982)

praspinal innervation of the lumbar cord, innervating the hindlimb motoneurons, comes from early arising tail-innervating reticulospinal and vestibulospinal axons, and that (2) in limbbud stages, a second wave of reticulospinal and vestibulospinal axons is aimed at the lumbar cord, thus innervating the lumbar, hindlimb-innervating motoneurons (ten Donkelaar et al. 1993; de Boer-van Huizen 1995). In one set of experiments, FDA was applied to the tail spinal cord of stage 48, 54, and 60 larvae. After survival times varying from 86–92 days in stage 48 experiments, 34–78 days in stage 54 experiments to 13 days in stage 60 experiments, metamorphosis was complete, and RDA was applied unilaterally to the lumbar spinal cord. Due to the long survival times in some of the stages 48 and 54 experiments no FDA could be demonstrated in the brain stem after metamorphosis. In those experiments in which enough FDA was found throughout the brain stem, double labeled cells were found in the interstitial nucleus of the flm, throughout the rhombencephalic reticular formation, in the inferior raphe nucleus, and in the vestibular nuclear complex. The number of RDA labeled cells by far exceeded that of the FDA labeled cells. Data were quantified for the interstitial nucleus of the flm, the middle and inferior reticular nuclei, the inferior raphe nucleus, and the large-celled vestibular nucleus (de Boer-van Huizen 1995). The percentage of double labeled (DL) cells varied from 25% in the inferior raphe nucleus to 70% in the interstitial nucleus, the vestibular nucleus, and the middle reticular nucleus in stage 48 experiments. In later stages, the number of DL cells in the interstitial nucleus, the vestibular nucleus, and the nucleus reticularis medius decreased drastically, through about 30%–50% in stage 54 experiments to 10%–20% in stage 60 experiments. The percentage of DL cells in the inferior raphe and reticular nuclei did not vary much and remained around 25%. These data suggest that brainstem neurons which innervate the tail cord before metamorphosis, do innervate the lumbar cord after metamorphosis. The decrease noted in later stages may be due to the retraction of supraspinal axons from the degenerating tail spinal cord. In a second set of experiments (ten Donkelaar et al. 1993), HRP was applied to the lumbar or tail spinal cord of stage 50 and 53 larvae. In these experiments, the number of labeled neurons in the reticular formation and the vestibular nuclear complex after lumbar cord injections by far exceeded that of tail injections. Similar observations were made using FDA and RDA. These data suggest the presence of a second wave of reticulospinal and vestibulospinal axons that is aimed at the lumbar cord, in line with birthday data (see Sect. 5.3). Mauthner cells maintain their lumbar projection in adult anurans (Will 1986, 1991; Davis and Farel 1990).

In *Rana catesbeiana*, Forehand and Farel (1982c) studied the development of brainstem projections to the lumbar spinal cord starting at Taylor and Kollros' (1946) stage I. As early as stage I, i.e., the first limbbud stage comparable to stage 46 in *Xenopus laevis* (see Table 1), labeled neurons were found throughout the reticular formation, in the raphe nucleus, in the ventral VIIIth nucleus, in the interstitial nucleus of the flm, and in the hypothalamus. The number of spinal projection neurons increased as the animals matured. No rubrospinal projection was noted, but hypothalamospinal projections were found much earlier than in *Xenopus laevis*.

Retinoic acid (RA) affects the organization of reticulospinal neurons in developing *Xenopus laevis* (Manns and Fritzsch 1991, 1992). This vitamin A derivative plays a role in the development of the vertebrate CNS (Maden and Holder 1991; Hofmann and Eichele 1994). In *X. laevis*, RA modulates *Hox*-gene expression, and causes an anterior truncation with an upregulation of *Hox*-gene expression in more anterior areas of the mesoderm and the neuroectoderm (Papalopulu et al. 1991; Sive and Cheng 1991). Low RA-concentrations led to a variable expression of additional Mauthner-like cells, whereas high concentrations resulted in their complete deletion. Apart from these effects on Mauthner cells, a consistent reduction of the degree of differentiation of the reticular formation with increased RA-concentrations was found (Manns and Fritzsch 1992). Normally, the reticular formation can be divided into mesencephalic, isthmal and three rhombencephalic nuclei, with a clear gap between the rhombencephalic and the mesencephalic groups. In the RA-treated animals, this gap disappears completely (Fig. 18). Moreover, in some cases vestibulospinal neurons disappeared completely. These data are consistent with the hypothesis that selective expression of developmental genes may be crucial for the patterning of specific areas of the hindbrain (Papalopulu et al. 1991).

In summary, HRP data show that interstitiospinal, reticulospinal and vestibulospinal fibers innervate spinal segments very early in development, prior to the development of the limb buds. The supraspinal innervation of the lumbar enlargement presumably starts shortly after the formation of the lateral column of motoneurons. Part of the supraspinal innervation of the lumbar cord comes from collaterals of early arising tail cord-innervating reticulospinal and vestibulospinal axons, another part from a second wave of reticulospinal and vestibulospinal axons directly aimed at the lumbar spinal cord. The anuran red nucleus projects spinalwards definitively later in development (Table 4). This *developmental sequence* parallels the changes observed in locomotor pattern. Until stage 58, locomotion (swimming) consists of coordinated, alternate contractions of the axial muscles on each side of the body. From stage 63 on, swimming is accomplished solely with the extremities. It is in this period that rubrospinal fibers innervate the lumbar spinal cord. In urodeles, tracer data on the development of descending supraspinal pathways are lacking.

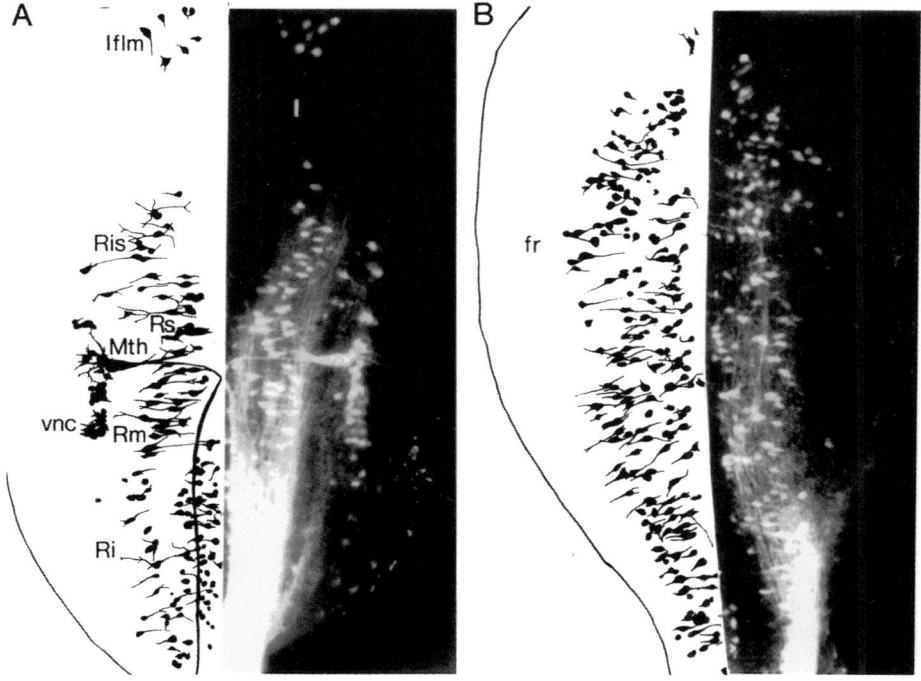

Fig. 18A,B. The effect of retinoic acid on the development of reticulospinal projections in a stage 48 *Xenopus laevis* tadpole. Photomicrographs of one half of a wholemounted brain and camera lucida drawings are shown for **A** a control and **B** a retinoic acid-treated animal. *fr*, formatio reticularis; *Iflm*, nucleus interstitialis of flm; *Ri, Ris, Rm, Rs*, nucleus reticularis inferior, isthmi, medius, and superior, respectively; *Mth*, Mauthner cell; *vnc*, vestibular nuclear complex. (After Manns and Fritzsch 1992)

45

Table 4. The development of descending spinal projections in anurans

Nuclei	Time of neuron origin (*Xenopus laevis*, peak production)[c,g] Nieuwkoop and Faber (1967) stages	Innervation of rostral cord in *X. laevis*[h] NF-stages	Innervation of tail cord in X. *laevis*[d] NF-stages	Innervation of lumbar cord in *Rana catesbeiana*[a] Taylor and Kollros (1946) stages
Reticular formation				
Nucleus reticularis inferior	10–13	28	37	I
Nucleus reticularis medius	10–13	28	37	I
Nucleus reticularis superior	10–13	35/36	37	I
Nucleus reticularis isthmi		43/44	43	I
Nucleus interstitialis of flm	10–13	30/31	39	I
Raphe nucleus	10–13	<35/36	41	I
Serotonergic projections[i]		32		
Vestibular nuclei				
Lateral vestibular nucleus	10–18	35/36	39	I
Medial and inferior vestibular nuclei	18–37/38	<50[e]		?
Mauthner cell	~10[c,j]	30/31	37	
Locus coeruleus				
Coeruleospinal neurons	?	43?	?	?
Noradrenergic projections[b]		41		
Red nucleus	?	48[f]		?
Hypothalamus	?	57[e]		I

[a] Forehand and Farel 1982c.
[b] González et al. 1994a.
[c] Lamborghini 1980.
[d] Nordlander et al. 1985.
[e] ten Donkelaar and de Boer-van Huizen 1982.
[f] ten Donkelaar et al. 1991.
[g] van Mier 1986.
[h] van Mier and ten Donkelaar 1984.
[i] van Mier et al. 1986.
[j] Vargas-Lizardi and Lyser 1974.

5.4.2
The Ingrowth of Reticulospinal, Monoaminergic and Rubrospinal Fibers

In the brain stem of *Xenopus laevis*, the first neurons differentiate at ventral and ventrolateral levels (Hartenstein 1993). These neurons send out descending axons that gather in a ventral longitudinal fascicle. Caudally, this bundle merges with the ventral fascicle of the spinal cord. The first regions to send axons into this ventral longitudinal tract are the reticular formation of the hindbrain and the interstitial nucleus of the flm (see Sect. 5.4.1.). In *X. laevis*, the *ingrowth* of reticulospinal axons into the spinal cord and the formation of contacts with primary motoneurons has been studied in the early swimming stage (van Mier and ten Donkelaar 1989). Starting at stage 25, the first primary motoneurons provided with dendrites are present in the rostral spinal cord (van Mier et al. 1985). *Reticulospinal fibers* directly contact primary motoneurons (Figs. 19, 20). The first ingrowth of reticulospinal axons was observed in the

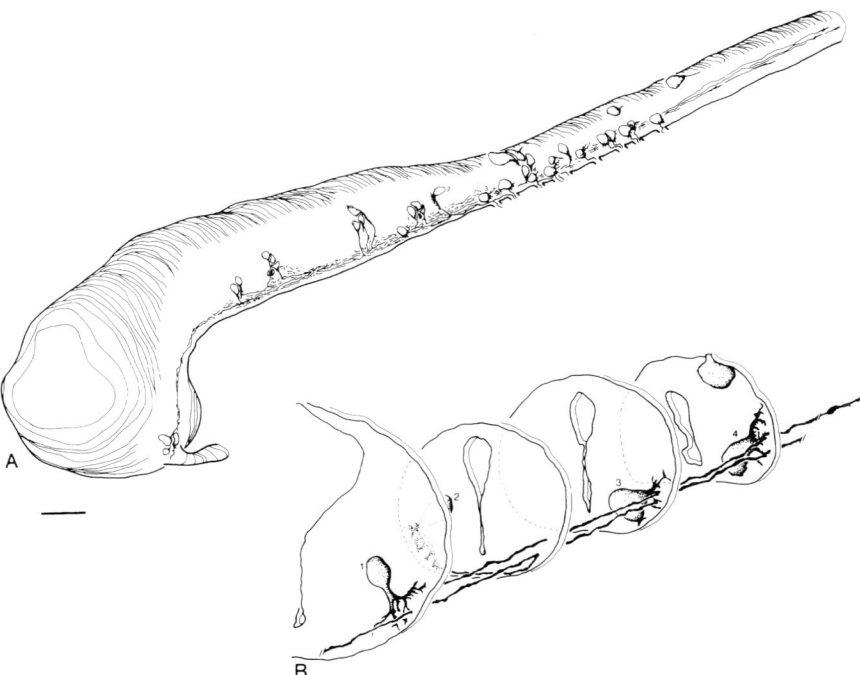

Fig. 19A,B. The ingrowth of reticulospinal axons into the spinal cord and the formation of synaptic contacts with primary motoneurons in *Xenopus laevis* shown in a serial reconstruction of a transversely sectioned CNS of a stage 33 embryo. HRP was applied to the caudal brainstem and to the 5th and 6th myotomes. **A** Lateral view of the reconstructed CNS. In the rostral spinal cord some primary motoneurons and Rohon-Beard cells can be seen. In the brain stem HRP labeled reticulospinal neurons projecting ipsilaterally or contralaterally to the spinal cord were found. **B** Four sections from this reconstruction showing an identified ipsilaterally (*1*) and a contralaterally (*2*) projecting reticulospinal neuron. The descending axons of these identified reticulospinal neurons could be traced to the spinal cord where they contacted two primary motoneurons (*3,4*) (From van Mier and ten Donkelaar 1989)

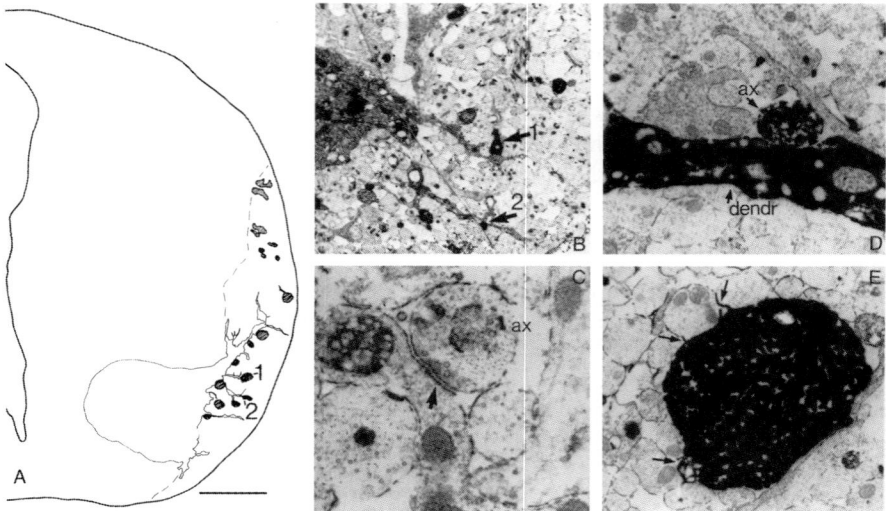

Fig. 20. A Transverse section of the spinal cord showing the primary motoneuron 3 of Fig. 19. The *hatched circles* are HRP labeled descending reticulospinal axons. The *numbered axons* correspond to the ipsilaterally (*1*) and contralaterally (*2*) projecting reticulospinal axons of Fig. 19. **B–E** Electronmicroscopical pictures showing **B** the sites where the descending reticulospinal axons contacted dendrites of primary motoneurons, **C** a chemical synapse between an axon and possibly a dendrite of a primary motoneuron, **D** a descending axon contacting a dendrite of a labeled primary motoneuron, and **E** a transverse section of a growth cone with some small protrusions (*arrows*) emerging from its surface. (From van Mier and ten Donkelaar 1989)

rostral spinal cord after application of HRP to the caudal brain stem of stage 27/28 embryos. Often one or two axons were observed far ahead ("pioneer fibers"). The descending fibers were usually observed in the ventral part of the marginal zone. By stage 32, near the time of the first rhythmic swimming (van Mier et al. 1989), many supraspinal axons reached the level of the 12th/13th myotome. Contacts between ingrowing reticulospinal axons and primary motoneurons were occasionally demonstrated in the spinal cord of stage 30 embryos, while at stage 33 more contact sites were observed. Reticulospinal fibers could be found contacting primary motoneurons anywhere on their dendritic trees, whereas the few descending axons of the contralaterally projecting reticulospinal neurons almost exclusively made contact on the ventrolateral dendrites close to the somata of the primary motoneurons (see Fig. 19). At stage 33, never more than one possible contact between a reticulospinal axon and a primary motoneuron could be observed, but the same descending axon could contact many primary motoneurons. Electronmicroscopically, these contacts were verified (van Mier and ten Donkelaar 1989). The contact sites appeared to be rather simple (Fig. 20) without axonal protrusions. Many labeled growth cones were observed descending along the spinal cord, usually close to the lateral border of the marginal zone.

Reticulospinal neurons participate in early swimming. During early locomotion, they have "motoneuron-like" activity (van Mier et al. 1989). Their activity during swimming shows a similarity to that reported for large reticulospinal neurons (Müller cells) in the lamprey, which are also active during each swimming cycle (Kasicki and

Grillner 1986). Lesion experiments (Roberts and Alford 1986) showed that the caudal brain stem and the rostral spinal segments have an important influence on the initial swimming frequency and the duration of the swimming period.

With transmitter immunocytochemistry early stages in the differentiation of a number of different neuron classes were observed in wholemount preparations of the developing *Xenopus* CNS. Roberts and co-workers studied the development of GABAergic and glycinergic neurons (Roberts et al. 1987, 1988). The first descending *serotonergic* axons appear in the rostral spinal cord at stage 32 (van Mier et al. 1986; see Fig. 21). At first, these fibers are found only in the dorsolateral part of the marginal zone, but later they are scattered over the marginal zone. It is likely that serotonergic projections to the motoneuron area arise before the serotonergic innervation of the

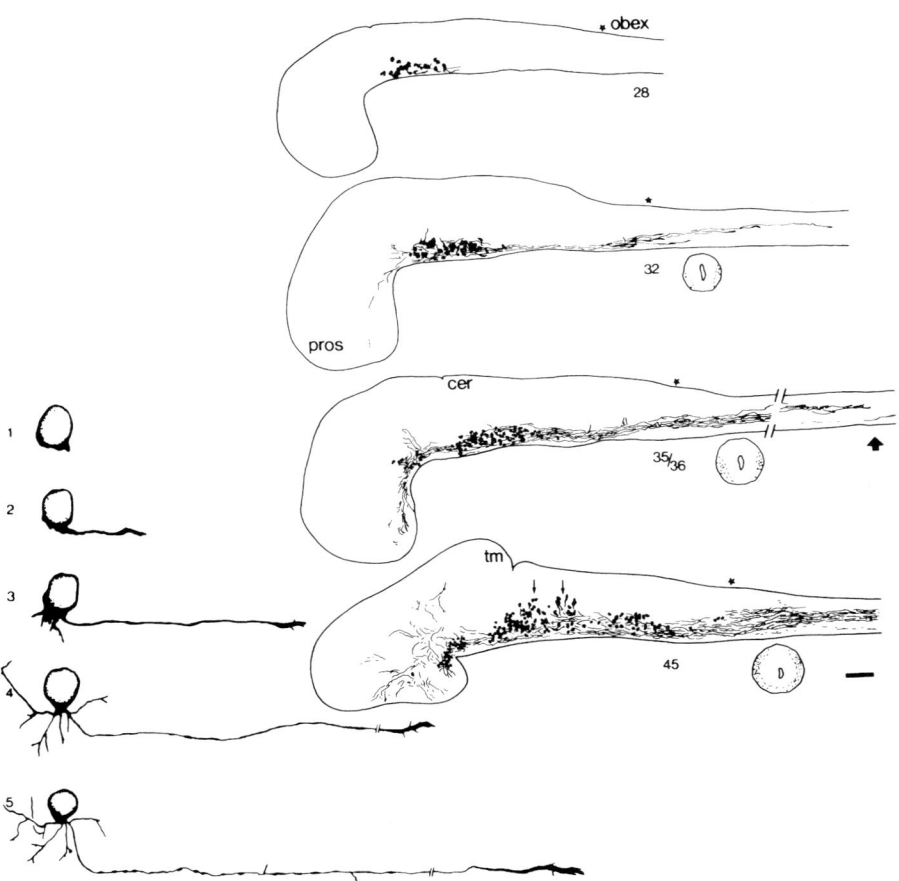

Fig. 21. Development of serotonergic neurons (*left panel*; see text for explanation) and the ingrowth of raphespinal fibers into the spinal cord (*right panel*) in *Xenopus laevis* shown in lateral views of the CNS. Transverse sections of the rostral spinal cord show the funicular trajectory of the serotonergic fibers. *Asterisk* indicates the obex. *cer*, cerebellum; *tm*, tectum mesencephali; *pros*, prosencephalon. (From van Mier et al. 1986)

dorsal horn. The development of individual raphespinal neurons can be characterized by five phases: (1) transmitter production and formation of an axonal protrusion, the initial outgrowth, at the ventral side of the young neuron; (2) formation of a fan-shaped growth cone bearing a few filopodia, after which axonal elongation takes place; (3) formation of the first dendrites at the rostroventral side of the soma; (4) maturation of the axon and the dendrites, i.e., in the thinner parts of the axon a few varicosities appear, the dendrites become thinner with the development of branches, and (5) development of axonal collaterals which extend into the spinal gray. In the axonal elongation phase, the serotonergic axons are exclusively found in a small bundle, suggesting that they may be guided by a pre-established pathway. Ingrowth of descending serotonergic fibers transforms the inflexible embryonic spinal locomotor network into a potentially more flexible adult-like one (Sillar et al. 1992, 1995).

González et al. (1994a,b) found tyrosine hydroxylase (TH) and dopamine (DA)-im-munoreactive fibers in the spinal cord of stage 41 *Xenopus laevis* tadpoles (Fig. 22). Remarkably, the appearance of longitudinal TH-immunoreactive fibers precedes that of TH-immunoreactive cells in the brain stem, since TH-immunoreactive neurons in the locus coeruleus also appear first at stage 41. In HRP studies possible *coeruleospi-nal projections* at the level of the nucleus reticularis isthmi were not found before stage 43 (van Mier and ten Donkelaar 1984; Nordlander et al. 1985). Double labeling experiments combining tract-tracing techniques with immunohistochemistry are necessary to solve this discrepancy. In *Rana perezi (ridibunda)*, TH-immunoreactive neurons in the locus coeruleus were first found at a comparable stage of development (see Table 1), i.e., at stage 31, just before the beginning of the larval period (González et al. 1994a,b). In urodeles, Sims (1977) described catecholaminergic projections to

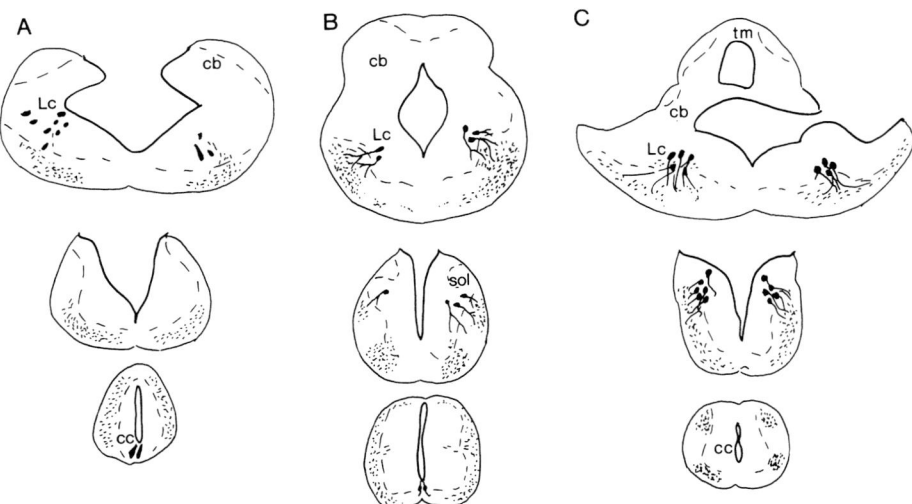

Fig. 22. Development of the catecholaminergic innervation of the spinal cord in **A,B** *Xenopus laevis* (stages 41 and 55, respectively), and **C** a stage 33a *Pleurodeles waltl* larva. *cb*, cerebellum; *cc*, central canal; *Lc*, locus coeruleus; *sol*, solitary tract; *tm*, tectum mesencephali. (Based on data by González et al. 1994a)

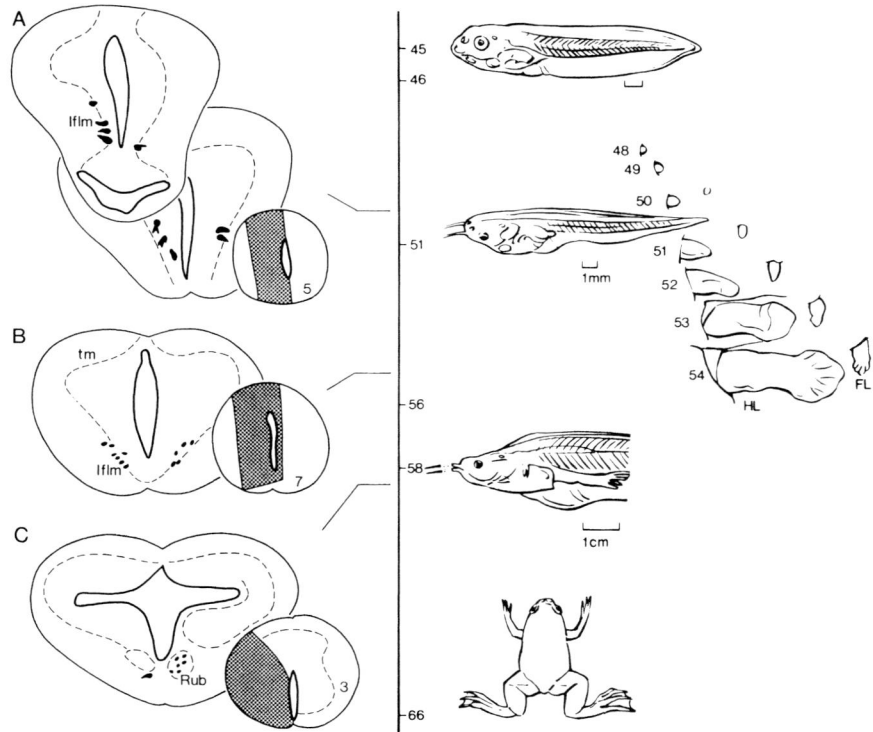

Fig. 23A–C. Development of rubrospinal projections in *Xenopus laevis*. HRP was applied to the spinal cord in stages 50 **A**, 55 **B**, and 58–59 **C**. *FL*, forelimb bud; *HL*, hindlimb bud; *Iflm*, interstitial nucleus of flm; *Rub*, nucleus ruber; *tm*, tectum mesencephali. (After ten Donkelaar and de Boer-van Huizen 1982)

the spinal cord in *Ambystoma mexicanum*, and González et al. (1995) in *Pleurodeles waltl* (staged after Gallien and Durocher 1957). In *P. waltl*, the locus coeruleus was observed at stage 33a (9 days). In the spinal cord, distinct tracts were found to descend through the dorsolateral part of the lateral funiculus and to terminate in the dorsal gray matter. On the basis of location, neurotransmitter content and efferent connections, the isthmic noradrenergic cell group of amphibians is considered homologous to the locus coeruleus of amniotes (Marín et al. 1996).

HRP studies (ten Donkelaar and de Boer-van Huizen 1982; ten Donkelaar 1988) suggested that *rubrospinal* neurons innervate the lumbar spinal cord rather late in development (at stage 58; see Fig. 23). Later HRP data (ten Donkelaar et al. 1991) showed that rubrospinal axons already invade the rostral spinal cord at stage 48 and reach the lumbar cord by about stage 55 (Fig. 24). The more ventral position, the smaller size of their somata and the crossing of their axons, makes it possible to distinguish rubrospinal neurons from the larger, more dorsally situated, ipsilaterally projecting interstitiospinal neurons. At stage 48, the rubrospinal neurons appear relatively immature with small, round to oval cell bodies and just the beginning of a dendritic tree. The rubrospinal tract neurons extensively increase their dendritic trees

51

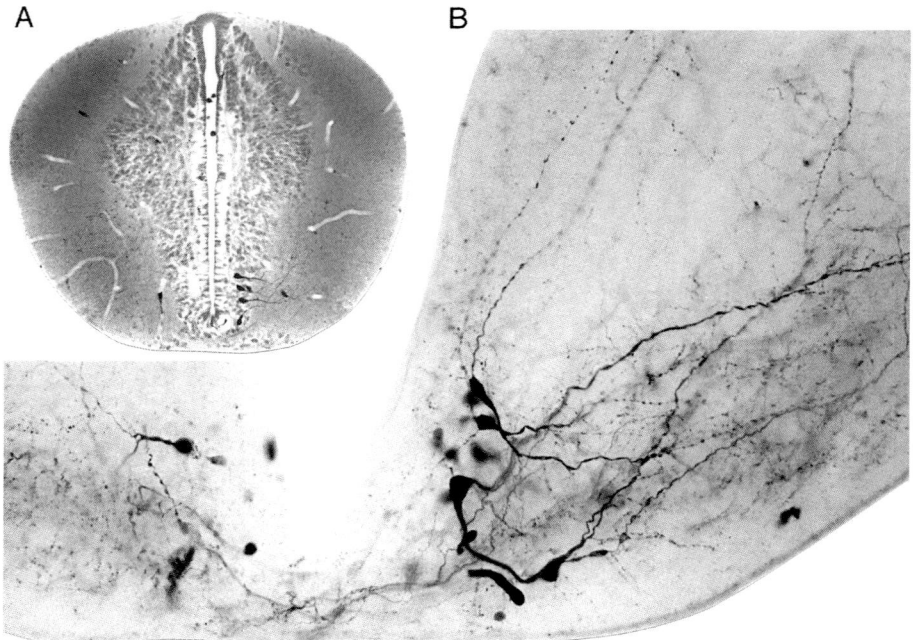

Fig. 24. Retrogradely labeled rubrospinal neurons in *Xenopus laevis* larvae after application of HRP at the spinomedullary border. **A** Photomicrograph showing the still undifferentiated rubrospinal neurons in stage 48. **B** Photomicrograph of differentiated rubrospinal neurons in stage 53. (From ten Donkelaar et al. 1991)

as shown by the labeled neurons in a stage 53 tadpole (Fig. 24). HRP data suggest that cerebellorubral projections arise around stage 49 in *X. laevis*, and are present well before the rubrospinal innervation of the spinal cord is complete (ten Donkelaar et al. 1991). In contrast, cerebellovestibular projections are formed somewhat earlier than the cerebellorubral projection (van der Linden and ten Donkelaar 1990), but innervate an already rather complete vestibulospinal system (van Mier and ten Donkelaar 1984; Nordlander et al. 1985).

5.4.3
Regenerative Capacity of Descending Brainstem Pathways

In contrast to urodeles, in juvenile and adult anurans no significant regeneration of descending supraspinal pathways occurs after transection of the spinal cord. In tadpoles subjected to cord transection, however, spinal cord continuity is readily restored (Sims 1962; Michel and Reier 1979; Forehand and Farel 1982a; Beattie et al. 1990). In *Rana catesbeiana*, tadpoles with spinal cord transections held through metamorphosis show, as juvenile frogs, restoration of lumbar projections from all brainstem regions that normally project to the lumbar spinal cord. Neither long ascending projections from dorsal root ganglion cells nor those arising from spinal neurons caudal to the transection were found to cross the transection site. If lesions were made early, i.e., up to stage VIII, however, dorsal column axons can grow across the transection as shown in *R. esculenta* (Clarke et al. 1986). Tadpoles, regardless of developmental stage, show only limited behavioral and anatomical recovery from the effects of spinal cord transection (Sims 1962; Forehand and Farel 1982a; Stehouwer 1986). After metamorphosis, however, juvenile frogs which had undergone spinal cord transection as tadpoles manifest obvious rostral control over caudal body regions.

In *Xenopus laevis*, Beattie et al. (1990) made complete transections of the thoracolumbar cord between stages 50 and 62. These extensive lesions were followed by gradual recovery of righting and coordinated swimming as animals metamorphosed into juveniles (stage 66). HRP tracing showed fibers crossing the lesion site and distributing to the lumbar enlargement. These fibers were traced from more rostral spinal segments as well as from the caudal brain stem. Juvenile toads of varying ages failed to demonstrate recovery of coordinated swimming or reconstitution of descending pathways. Immunohistochemical studies showed that serotonergic fibers were included in the population of axons that rapidly crossed the lesion after transection at metamorphic stages. It was suggested that both metamorphosis-related hormonal changes and axon substrate pathways may affect the regenerative response in the CNS of *X. laevis*. Using dextran amines, ten Donkelaar et al. (1993) applied a sequential double labeling technique. A complete transection of the thoracic cord was made at stage 58. The dextran amine FDA was applied to the rostral stump of the lesioned cord. After metamorphosis, RDA was applied to the lumbar enlargement. Three populations of neurons were found: (1) FDA labeled neurons, retrogradely filled from the stump of the lesioned cord; (2) RDA labeled neurons, retrogradely labeled from the reconnected lumbar spinal cord, and (3) double labeled cells, i.e., neurons that projected to the lumbar spinal cord or beyond at stage 58 and reinnervated the lumbar spinal cord after the lesion. These experiments show that like in urodeles (Davis et al. 1989a) neurons throughout the brain stem have found their way to the lumbar enlargement. A limited number of double labeled cells were found in the reticular formation, the serotonergic inferior raphe nucleus, in the vestibular nuclear complex, and in the interstitial nucleus of the flm. Neurons only labeled by RDA presumably project to the lumbar cord late in development, and have found their way via the reconnected spinal cord.

Data by Lang et al. (1995) revealed a striking difference in substrate properties of CNS myelin and oligodendrocytes derived from *Xenopus* hindbrain/spinal cord or optic nerve/tectum. *Xenopus* spinal cord myelin and oligodendrocytes possess nonpermissive substrate properties for growing axons in vitro to which mammalian-like neurite growth inhibitors contribute to a considerable degree. In contrast, oligodendrocytes and myelin derived from *Xenopus* optic nerve and tectum mesencephali are growth-permissive substances, and neurite growth inhibitors were not detectable. This difference in substrate properties and expression of neurite growth inhibitors correlates with the failure of axonal regeneration in the adult *Xenopus* spinal cord (Beattie et al. 1990) and the success of retinal axon regrowth in the visual system (see Gaze 1970; Jacobson 1991). Outgrowth assays showed that *Xenopus* optic nerve/tectum myelin allows axon outgrowth to a similar extent as goldfish CNS myelin (Bastmeyer et al. 1991). *Xenopus* spinal cord myelin, however, is a nonpermissive substrate in that outgrowth of axons is as poor as on rat CNS myelin (Bastmeyer et al. 1991).

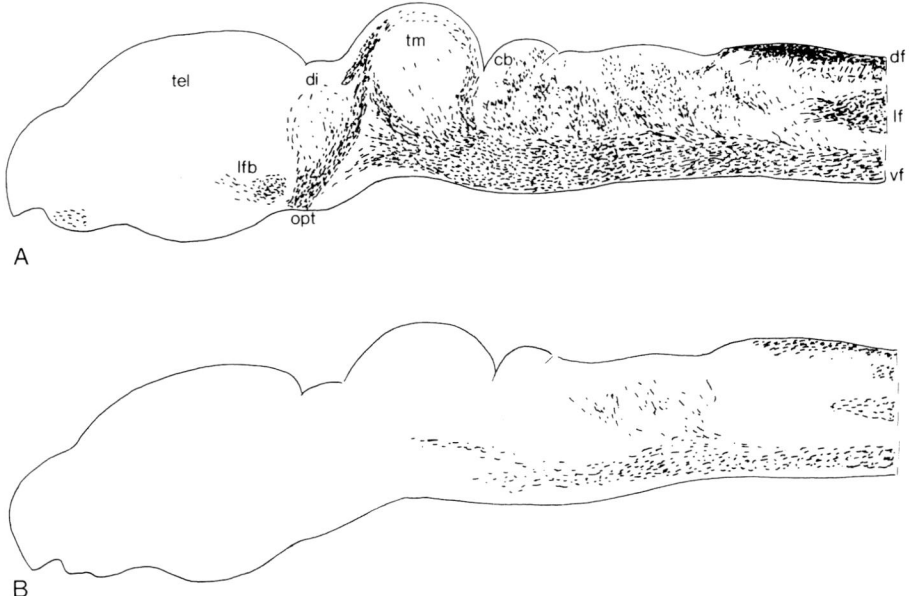

Fig. 25. A Myelinated tracts in the CNS of an adult *Xenopus laevis* demonstrated by a PLP antibody. **B** The presence of IN-1 immunoreactivity in the adult *X. laevis* CNS. *cb*, cerebellum; *di*, diencephalon; *df*, dorsal funiculus; *lf*, lateral funiculus; *lfb*, lateral forebrain bundle; *opt*, optic tract; *tel*, telencephalon; *tm*, tectum mesencephali; *vf*, ventral funiculus. (After Lang et al. 1995)

Lang et al. (1995) also applied the monoclonal antibody (Mab) IN-1 to visualize the distribution of mammalian-like neurite growth inhibitors in CNS myelin of *X. laevis* (Fig. 25). Tadpoles and metamorphosing animals were included to determine the onset and extent of myelination and appearance of IN-1 staining during development. Myelinated tracts were identified by antibodies against the myelin proteins PLP and MBP. The degree of myelination in premetamorphic tadpoles was low. In the spinal cord of a stage 51 tadpole, PLP and MBP staining was restricted to a narrow superficial zone in the ventral spinal cord corresponding to the position of reticulospinal and Mauthner axons. At the beginning of the metamorphic climax (stage 61), myelinated fibers were found throughout the ventral and lateral funiculi as well as in the dorsal funiculus. At stage 61, staining with Mab IN-1 became just detectable in the dorsal funiculus. In the adult CNS of *Xenopus laevis*, myelination was present in all spinal funiculi, in brainstem pathways, in the tectum mesencephali, in the optic nerve and tract and in the lateral forebrain bundle (Fig. 25A). Mab IN-1 labeling was restricted to myelinated areas of the spinal cord and brain stem (Fig. 25B). These data indicate that myelin from adult *X. laevis* optic nerve/tectum has no detectable amount of mammalian-like myelin-associated inhibitors defined by Mab IN-1, whereas myelin from adult *Xenopus* spinal cord appears to possess them in a considerable degree.

6 Development of Descending Supraspinal Pathways in Birds

6.1
Some Notes on the Development of the Avian Brain Stem and Spinal Cord

The development of the avian spinal cord is not very advanced at the time the earliest descending projections have reached the cord. In the chicken cervical cord, spinal motoneurons are first born at about 28 h of development comparable to Hamburger and Hamilton's (HH) stage 8, those in the brachial cord around 52 h (stage HH15), and those in the lumbar cord around 56 h or stage HH17 (McConnell and Sechrist 1980). Motor projections from cervical, brachial and lumbar levels of the cord first appear on day 3 (Windle and Austin 1936; Hollyday 1983), whereas sensory projections to the cervical and lumbar levels first appear on days 3 and 6, respectively (Windle and Austin 1936; Davis et al. 1989b). Davis et al. (1989b) showed that primary afferent fibers reach the lumbosacral cord at stage HH24 (E4), extend into the dorsal funiculus rostrally and caudally for over 24 h, and invade the gray matter at HH28 (E6). Dorsal root fibers extend into the vicinity of motoneuron dendrites by HH32 (E7.5), i.e., about the time that reflexes can first be elicited. Dense projections to the dorsal layers of the lumbosacral spinal cord, presumably from cutaneous afferent fibers, appear at about HH39 (E13), when the segmental pattern begins to resemble the mature pattern. The early development of spinal interneurons was extensively studied (see Nornes et al. 1980a,b; Oppenheim et al. 1988; Shiga and Oppenheim 1991; Yaginuma et al. 1991, 1994). By E6 spinal interneurons can be classified into five categories depending on their cell location, and the laterality and rostrocaudal direction of their axonal projections (Oppenheim et al. 1988; Yaginuma et al. 1994). Intersegmental projections between brachial and lumbar regions are established by interneurons as early as E4–4.5 (Oppenheim et al. 1988), and the first synapses between interneurons and motoneurons are formed by E4.5–5 (Shiga et al. 1991). Descending supraspinal fibers reach the ventral funiculus at brachial and lumbar levels by E4 and E5, respectively (Okado and Oppenheim 1985).

Using silver impregnation techniques, Tello (1923) and Windle and Austin (1936) reported that reticular neurons were the first to begin neurofibrillar development, doing so around 40 h of incubation (about HH11). At this stage of development both reticulospinal fibers and the fasciculus longitudinalis medialis can be distinguished (Fig. 26). With the neurite-specific G4-antibody, Weikert and colleagues (1990) found labeling of the reticulospinal tract at HH11, directly followed by the flm. Chédotal and

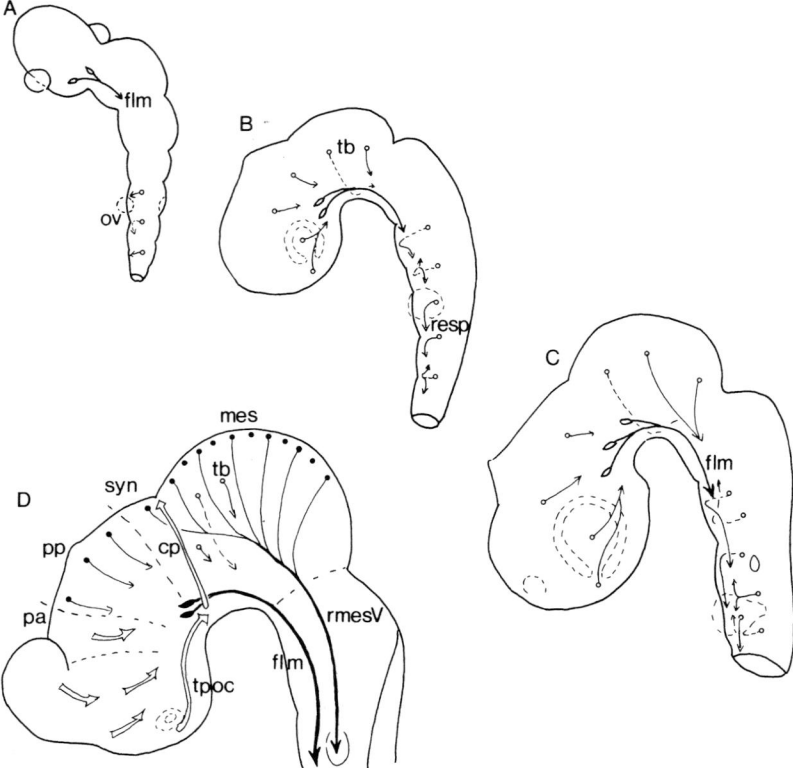

Fig. 26A–D. Initial tract formation in the avian brain as found in silver impregnation studies (**A–C**) and by applying specific antibodies (**D**). **A** About HH11/12, **B** about HH13, **C** about HH17, **D** the organization of the early axonal tracts in the chicken forebrain and brain stem labeled by BEN (*black dots* and *arrows*) up to HH20. The early developing tracts of the posterior and postoptic commissures as well as neurons in the telencephalon and anterior parencephalon do not express BEN (*open arrows*). *cp*, commissura posterior; *flm*, fasciculus longitudinalis medialis; *mes*, mesencephalon; *ov*, optic vesicle; *pa*, *pp*, anterior and posterior parencephalon, respectively; *resp*, reticulospinal neurons; *rmesV*, mesencephalic root of the trigeminal nerve; *syn*, synencephalon; *tb*, tectobulbar neurons; *tpoc*, tract of the postoptic commissure. (**A–C** after Windle and Austin 1936; **D** based on Chédotal et al. 1995)

co-workers (1995) studied the expression of β-tubulin and an immunoglobulin (BEN) during early stages of the first axonal tract formation in chicken embryos of stages 12–22. In line with the classical silver impregnation data (Tello 1923; Windle and Austin 1936), they found that the flm is the first tract to form in the forebrain of the HH12–13 chicken embryo (Fig. 26). The neurons of the interstitial nucleus of the flm become positive for β-tubulin around stage HH12.

6.2
Birthday Studies in Chickens

Reticular neurons are among the first born in the entire chicken neuraxis, their generation beginning near the end of the first embryonic day, i.e., at stage HH5 (McConnell and Sechrist 1980; Sechrist and Bronner-Fraser 1991). More rostral brainstem neurons are born shortly after those in the hindbrain. The neurons of the interstitial nucleus of the flm form an exception. They appear by stage HH7 at the forebrain-midbrain junction preceding neurons of the more caudal pontine reticular formation. The first vestibular neurons are born about a day later (McConnell and Sechrist 1980). In general, the expression of neurofilament immunoreactivity by early neuroblasts precedes axon outgrowth by several stages (Sechrist and Bronner-Fraser 1991). Neurofilament expression in reticular neuroblasts first occurs by the 7-somite stage (HH9), followed by axon outgrowth by the 15-somite stage (HH11–12). At the time of initial neurofilament expression, reticular neurons are bipolar in a columnar epithelial configuration and do not have an axon. In the interstitial nucleus of the flm, neuroblasts first express neurofilament at stage 10 and elicit axons shortly after reticular neurons in the hindbrain. Thus, reticular precursors cease mitosis prior to neural tube closure and express neuron-specific proteins shortly after closure (Sechrist and Bronner-Fraser 1991).

6.3
Tract-Tracing Data in the Chicken Embryo

6.3.1
Retrograde Tracer Data

Okado and Oppenheim (1985) have labeled reticulospinal neurons from the brachial cord, i.e., the cervical enlargement, as early as day 4 and from the lumbar cord as early as day 5, with vestibulospinal projections lagging by a half to one day (Table 5). Long propriospinal fibers from the brachial cord innervate the lumbar cord by E5, those from the cervical cord by E5.5. Ascending spinal projections from the lumbar cord reach the mesencephalon by E5. Thus, the earliest descending supraspinal, propriospinal, and ascending projections appear to be formed around the same time (see also Nornes et al. 1980a,b). At E7, the first lumbar spinocerebellar fibers reach the cerebellar plate, and at E8, the majority of the spinocerebellar projections (Lakke et al. 1985, 1986; Okado et al. 1987). Interstitiospinal fibers reach cervical levels by E5, and the lumbar cord at E5.5. Rubrospinal neurons are first labeled from the cervical cord at E7, and from the lumbar cord at E8. Raphespinal projections from both pontine and medullary levels reach the cervical cord at E5, but the lumbar cord at different times: pontine raphespinal fibers reach lumbar levels by E5.5, but medullary raphespinal fibers by E8. Coeruleospinal fibers innervate the cervical cord at E6 and the lumbar cord at E8. Remarkably, the first TH-immunoreactive neurons in the locus coeruleus were not observed before E8 (Puelles and Medina 1994). Hypothalamospinal projections from an unidentified hypothalamic nucleus reach the lumbar cord at E5.5 (Okado and Oppenheim 1985). Between E6 and E10, a large number of labeled ventral hypothalamic neurons, located just dorsal to the optic chiasm, were found labeled

Table 5. The development of descending supraspinal projections in chickens

Nuclei	Time of neuron origin[b]	Innervation of cervical enlargement[c]	Innervation of lumbar cord[c]
Reticular formation			
Medullary	20 h (HH5)	E4 (HH23)	E5
Pontine	24–28 h (HH7/8)	E5 (HH26)	E5.5–7
Interstitial nucleus flm	20–24 h (HH7)	E5	E5.5 (HH27/28)
Raphe nuclei	~24 h (HH7)	E5	E5.5(?)-8
Serotonergic projections[d]		E6 (HH28/29)	E8
Vestibular nuclei			
Lateral vestibular nucleus	32 h? (HH9/10)	E5	E5.5
Medial and inferior nuclei		E8 (HH34)[a]	
Locus coeruleus			
Coeruleospinal projections	24 h (HH7)	E6	E8
Red nucleus	28 h (HH8)	E7 (HH30)	E8
Hypothalamus			
Paraventricular nucleus		E8	E12 (HH38)

[a] Glover and Petursdottir 1991.
[b] McConnell and Sechrist 1980.
[c] Okado and Oppenheim 1985.
[d] Okado et al. 1992.

after cervical HRP injections. These cells may represent transient projections, since no labeled cells were found in this region in the newly-hatched chicken. Projections from the paraventricular nucleus reach the cervical cord at E8, and the lumbar cord by E12.

In the chicken embryo, Glover and Petursdottir (1988, 1991) studied the regional specificity of developing reticulospinal and vestibulospinal projections. In in vitro preparations (see Glover et al. 1986), brainstem cell populations were labeled from 3 days of embryonic development (about HH18), when the first reticulospinal fibers reach the cervical cord (Fig. 27), to 11 days of development, when both systems have acquired a relatively mature organization. A striking feature at all stages was the spatial segregation of many neuron groups projecting along different trajectories. This segregation suggests that the choice of projection pathway by many brainstem neurons is in some way linked to cell position. Other clusters, however, overlap extensively linking a particular region to more than one pathway. At least some of this intermingling appears to occur subsequent to the initial establishment of axon projection systems. The reticulospinal and vestibulospinal axons project along two major tracts: a *medial bundle*, originally termed the common longitudinal bundle (Windle and Austin 1936), and a *lateral bundle*. The medial tract corresponds to the *medial longitudinal fascicle*, whereas the lateral bundle forms the *lateral vestibulospinal tract* of older stages (Glover and Petursdottir 1988). The first reticulospinal axons all enter the medial longitudinal fascicle (Fig. 27A). At early stages (Fig. 27B), the lateral bundle contained only ascending spinocerebellar fibers, but by day 4 (HH24–26) a few

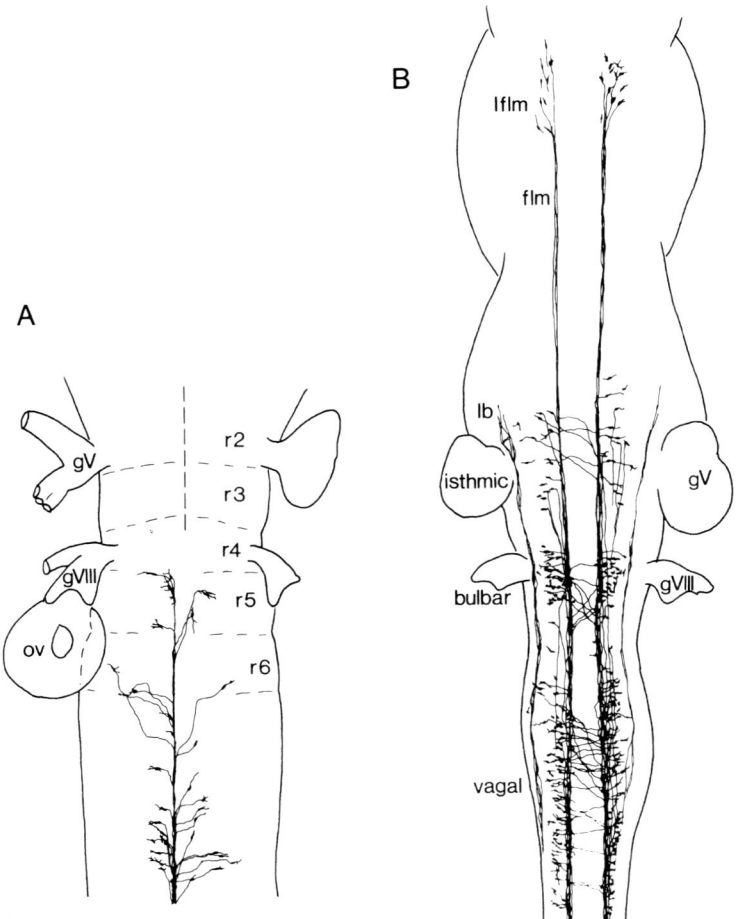

Fig. 27A,B. The earliest developing reticulospinal projections in the chicken embryo. **A** Stage HH19 (E3), **B** stage HH20.5 (E3). In **A** HRP was injected unilaterally into the cervical spinal cord, in **B** bilaterally. *bulbar, isthmic, vagal,* embryonic clusters of corresponding reticulospinal cell populations; *gV, gVIII,* ganglia of trigeminal and octaval nerves, respectively; *flm,* fasciculus longitudinalis medialis; *Iflm,* interstitial nucleus of flm; *lb,* lateral bundle; *ov,* otic vesicle; *r2–r6,* rhombomeres. (After Glover and Petursdottir 1991)

vestibulospinal axons as well as some spinal trigeminal axons were found to enter the lateral bundle (Glover and Petursdottir 1991). At later stages, the three axon types of the lateral bundle – spinocerebellar, vestibulospinal, and spinal trigeminal – run in widely separate tracts. Medial vestibulospinal projections (Fig. 28) were found to pass via the medial longitudinal fascicle (Glover and Petursdottir 1988, 1991). Late day 8 (HH34) was the earliest day the cells of origin of the ipsilateral and contralateral vestibulospinal tracts could be unambiguously labeled.

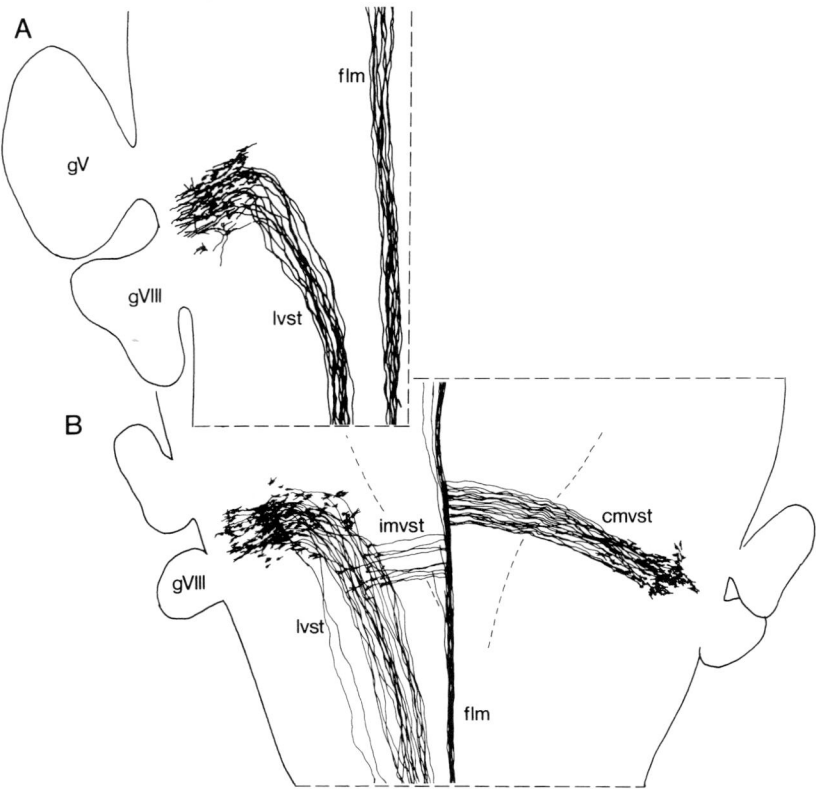

Fig. 28. The development of vestibulospinal projections at day 7 (**A**) and day 9 (**B**). *gV, gVIII*, ganglia of trigeminal and octaval nerves; *flm*, fasciculus longitudinalis medialis; *imvst, cmvst*, ipsilateral and contralateral medial vestibulospinal tracts, respectively; *lvst*, lateral vestibulospinal tract. (After Glover and Petursdottir 1991)

By 11 days of development, the major nuclei in the brain stem and spinal cord are formed and many of them have established synaptic connections. Retrograde tracing studies (Glover and Petursdottir 1988, Glover 1993) showed that at E11 the organization of descending supraspinal projections is rather similar to that seen in the adult chicken. Mesencephalic, pontine, and medullary reticulospinal axons all pass via the medial longitudinal fascicle (Fig. 29), raphespinal projections in a more lateral position, whereas three clusters of vestibulospinal neurons have their own characteristic pathway. A developmental time sequence was found in the appearance of projections to the spinal cord from the various brainstem clusters (Glover and Petursdottir 1991; Glover 1993). The first brainstem neurons to project to the spinal cord are the medullary reticulospinal (on day 3), followed shortly by the pontine reticulospinal and interstitiospinal projections. On day 4, a prominent contralateral pontine reticulospinal projection has developed. The raphespinal and lateral vestibulospinal projections also appear about this time. The contralateral pontine reticulospinal projection is a

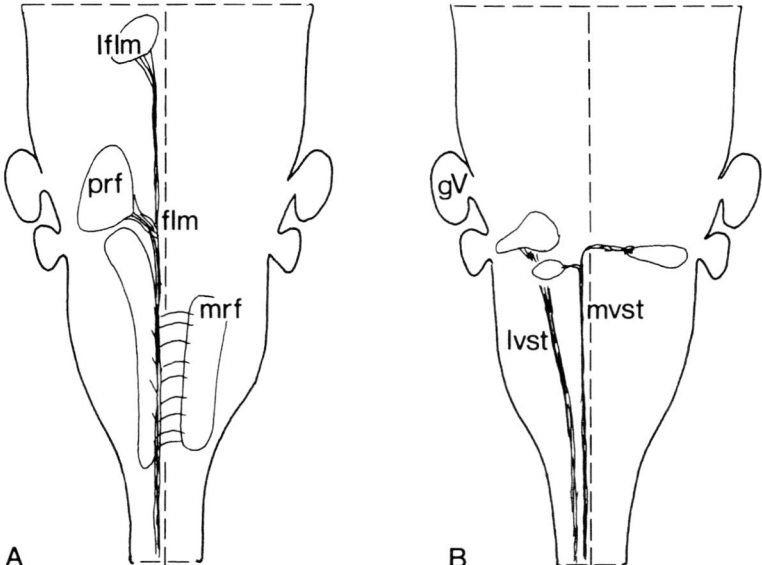

Fig. 29A,B. Summary of the organization of brainstem projections in the 11 day chicken embryo. **A** the interstitiospinal and reticulospinal projections **B** the vestibulospinal projections. *gV*, trigeminal ganglion; *flm*, fasciculus longitudinalis medialis; *Iflm*, interstitial nucleus of flm; *lvst*, *mvst*, lateral and medial vestibulospinal tracts, respectively; *mrf*, *prf*, medullary and pontine parts of reticular formation, respectively (After Glover 1993)

transient one: it is retracted during the next 3–4 days. The last elements to be added are the medial vestibulospinal projections.

The domains of the vestibulospinal (and vestibulo-ocular) projections are delineated rostrally and caudally by rhombomere boundaries, while the domains of these and the reticulospinal clusters are delineated mediolaterally by covert longitudinal boundaries (Glover 1993). Thus, the brainstem-spinal and vestibulo-ocular clusters appear to arise through the segmentation of parallel longitudinal cell columns, and the pathway choice exhibited by a neuron is defined by the column and the segment in which it is situated. By combining homotypic grafts of single quail rhombomeres (either r3, r4 or r5) with in vitro retrograde labeling at later stages of quail/chick chimeras, Díaz et al. (1998) showed that of the three vestibulospinal groups defined by axonal trajectory, one is constrained mainly to r4, one lies exclusively in r5, and one lies in a mediolaterally intermediate part of both r5 and r6. Thus, the trajectory-defined vestibulospinal cell populations correlate strongly with one or two specific rhombomeres.

6.3.2
The Ingrowth and Synaptogenesis
of Reticulospinal and Monoaminergic Fibers

Shiga and co-workers (Shiga et al. 1991) studied the axonal projections and synaptogenesis by supraspinal descending neurons in the spinal cord of the chicken embryo. In in vitro experiments they applied HRP to the boundary of the brainstem and spinal cord at E3.5 to E7. By E4.5, supraspinal fibers were found to descend to brachial segments through the ventral and lateral funiculi, and to enter the gray matter from the lateral funiculus by E6.5. Therefore, there is a *delay* of about two days between the arrival of supraspinal fibers in the white matter and the penetration of the gray matter. On E6, supraspinal descending fibers establish axodendritic synapses in the white matter. Therefore, there is also a delay of about one and a half days between the arrival of supraspinal fibers at a given spinal segment and the onset of synapse formation. Synapse formation from propriospinal sources, i.e., from spinal interneurons precedes that from supraspinal descending axons (Oppenheim et al. 1988; Yaginuma et al. 1991, 1994).

Okado and co-workers (Sako et al. 1986a, b; Okado et al. 1992) studied the development of the serotonergic system in the brain and spinal cord of the chicken. Serotonin-immunoreactive neurons appear in the rhombencephalon as early as E4. Two clusters of serotonergic neurons were found, one in the rostral part of the rhombencephalon corresponding to the dorsal raphe nuclei, the other located in the caudal part of the rhombencephalon corresponding to the medullary raphe nuclei of adult chickens. Serotonin-positive fibers were first recognized in the spinal cord as early as E6 and E8 in the marginal zones of the cervical and lumbar enlargements, respectively (Fig. 30). There was a waiting period of a few days before they penetrated into the gray matter. Terminal arborization of serotonergic fibers started around E16, and was maximized within one week of hatching. Thereafter, the density of serotonergic fibers decreased throughout the spinal cord. The density of serotonin was found to increase linearly from E6 to hatching, and reached its maximum value one week posthatching. Following the transient increase serotonergic fibers were eliminated from the neuropil. A particular high density of serotonin-positive fibers was found to the motoneurons of the extensor muscles of the hip joint. Okado's data show that there are four stages in the development of the projections from the raphe nuclei to the spinal motoneurons: 1) pathway formation in the marginal layer at early embryonic days; 2) terminal formation in the ventral horn in the period around hatching; 3) transient increase of serotonergic fibers around one week posthatching; and 4) localization of serotonergic fibers in specific motoneuron pools.

Tracing studies in chicken embryos demonstrate that descending reticulospinal pathways extend to all levels of the spinal cord during development (Okado and Oppenheim 1985; Shiga et al. 1991), and form synaptic connections within 1–2 days of reaching their target (Shiga et al. 1991). Behavioral studies show that although neurogenic axial motor activity is present in the embryo before the extension of descending pathways, the first detectable limb movements begin only after the first reticulospinal pathways have extended into the spinal cord (Bekoff 1981, 1992). Sholomenko and O'Donovan (1995) studied the development of the descending brainstem pathways to the lumbosacral cord in an isolated brain stem-spinal cord preparation. At E6, lumbosacral motor activity could be evoked by brainstem electrical stimulation. At E5,

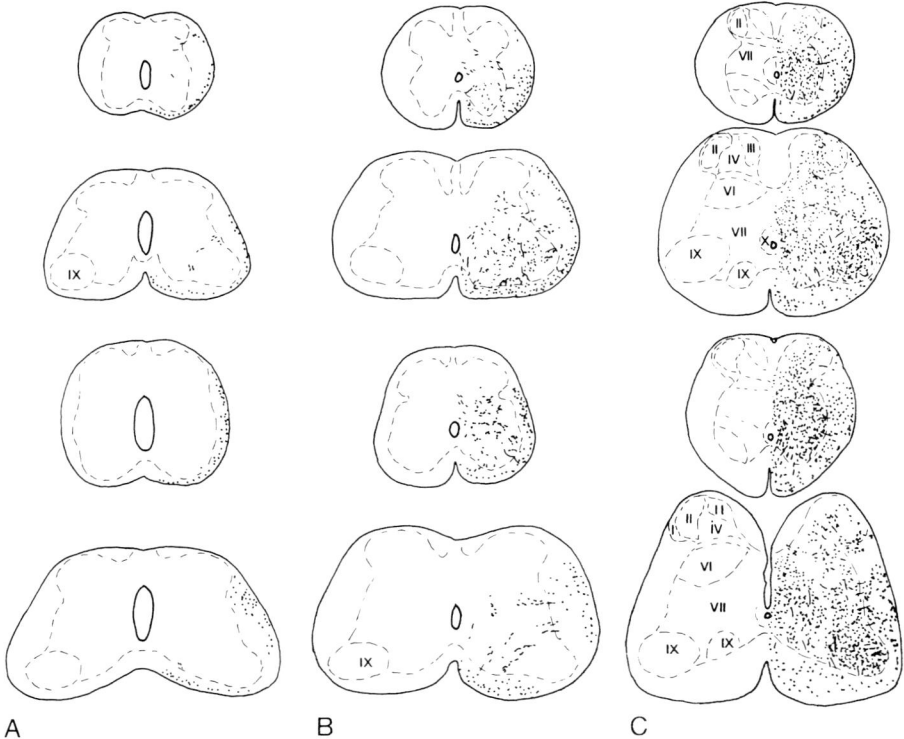

Fig. 30A–C. The development of the serotonergic projection to the spinal cord in the chicken embryo. **A** E8, **B** E10, **C** E16. The *first*, *second*, *third* and *fourth rows* represent midcervical, brachial, midthoracic and lumbosacral levels, respectively. *I–X* indicate the subdivision of the spinal gray matter. (After Sako et al. 1986b)

brainstem stimulation did not evoke such motor activity. Lesion and electrophysiological studies indicated that the activation of lumbosacral networks (see O'Donovan and Landmesser 1987; O'Donovan et al. 1992, 1994) is mediated by axons traveling in the ventral cord. By combining brainstem stimulation with retrograde tracing from the spinal cord, it was found that the brainstem regions from which spinal activity could be evoked lie in the reticular formation close to neurons that send reticulospinal axons to the lumbosacral cord. The cells of origin of these descending pathways were found in the ventral pontine and medullary reticular formation, a region that is the source of reticulospinal neurons important for motor activity in adult birds (Sholomenko and Steeves 1987; Steeves et al. 1987).

6.3.3
Regenerative Capacity of Descending Brainstem Pathways

The embryonic chicken is capable of axonal repair after complete transection of the thoracic spinal cord (Shimizu et al. 1990; Hasan et al. 1991, 1993; Steeves et al. 1994). Shimizu et al. (1990) studied the anatomical and functional recovery following spinal cord transection in the chicken embryo. The first reticulospinal projections descend to the lumbar cord by E5, and the descent of all descending brainstem projections to lumbar levels is essentially complete between E10 and E12. Transection of the cord on E2, when neurogenesis is still occurring and descending brainstem pathways have not been formed, produced no noticeable behavioral or anatomical deficits. Embryos hatched on their own and were behaviorally indistinguishable from controls. Similar observations were made at E5, when neurogenesis is complete and descending and ascending spinal pathways are developing. Embryos transected on E5 were able to hatch and could stand and locomote posthatching in a manner that was indistinguishable from controls. Following transections on E10, anatomical recovery of the spinal cord at the site of the lesion was not quite as complete as on E5, but continuity was restored, axons projected across this region, and rostral spinal and brainstem neurons could be retrogradely labeled after HRP injections caudal to the lesion. Repair of the spinal cord on E15 was considerably less complete compared to embryos transected earlier. Labeled brainstem neurons were never observed following transections on E15. Therefore, successful anatomical and functional recovery following a complete lesion of the spinal cord appeared to be limited to the period between E2 and E10.

Hasan and co-workers (Hasan et al. 1991, 1993; Steeves et al. 1994) further studied the functional repair of the transected spinal cord in embryonic chickens. Transections of the thoracic cord and sham operations were made on embryos from E3 to E14. After a recovery period of 5–18 days, the extent of the anatomical repair was studied by injecting a retrograde tracer into the upper lumbar cord. Anatomical recovery appeared to be complete for embryos transected as late as E12. Thoracic cord transections made on E13 or E14 resulted in reduced labeling of most brainstem nuclei known to project to the spinal cord. Functional recovery was assessed by behavioral observations and by focal electrical stimulation of brainstem locomotor regions. Locomotion in hatchling chickens or brainstem evoked locomotion in animals transected on or before E12 was indistinguishable from that observed in control chickens, indicating that complete functional recovery had occurred. Regeneration of previously axotomized projections may account for some of the observed anatomical and functional repair of descending brainstem pathways. By using two different retrograde fluorescent tracers (FDA and RDA), the first injected into the lumbar cord prior to thoracic transection and the second after the transection, Hasan et al. (1993) showed that spinal cord repair in the permissive period is due to true axonal regeneration. Reticulospinal pathways probably subserve a significant part of the recovery of motor function, since the preservation of these pathways is essential for voluntary walking in birds (Sholomenko and Steeves 1987; Steeves et al. 1987). The subsequent return of voluntary function results from reconnection of spinal CPGs with supraspinal axons which established connections with local circuits after extending 1–3 segments through a transection (Sholomenko and Delaney 1998). Thus, propriospinal axons were responsible for transferring input from descending excitatory axons which had regrown a short distance across a transection to more caudal segments.

Myelin within the embryonic chicken spinal cord inhibits functional regeneration of axotomized brainstem-spinal projections (Keirstead et al. 1992). The development of myelin in the chicken spinal cord begins on E13 and is completed prior to hatching (Macklin and Weill 1985; Keirstead et al. 1992). The functional regeneration of descending spinal axons is therefore restricted to the period prior to the developmental onset of myelination. The permissive period can be extended beyond E13 by preventing or disrupting myelination with specific immunological suppression. Blocking the onset of myelination with myelin-associated inhibitors of axonal growth at E9-E12 resulted in a delay of the onset of myelination until E17 (Keirstead et al. 1995). A subsequent transection of the spinal cord as late as E15 was followed by regeneration of descending pathways and functional recovery.

7 Development of Descending Supraspinal Pathways In Opossums

7.1
Some Notes on the Development of the CNS in Opossums

The external morphology of pouch young of the North American opossum was described by McCrady (1938) and Ulinski (1971). Prenatal opossums are shown in Fig. 31. Brain development of the North American opossum goes through three periods (Ulinski 1971): (1) an *embryonic* period extending from conception to the third postnatal day (P3) – by P3 the brain has undergone a longitudinal compression which gives it the shape of an adult brain; (2) a period of *regional differentiation* from P4 to about P60 during which all of the macroscopically visible adult brain structures sequentially appear within each region; and (3) a period of *maturation* from about P60 to weaning.

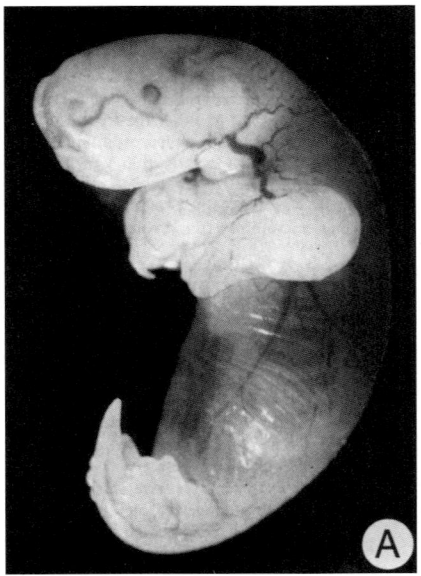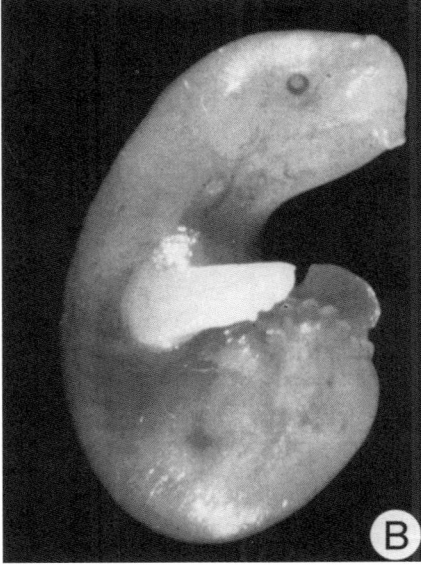

Fig. 31A,B. Photomicrographs of newborn opossums. **A** The North American opossum, *Didelphis virginiana*. **B** The Brazilian gray short-tailed opossum, *Monodelphis domestica*. (Photomicrographs kindly provided by Dr. George F. Martin)

In the two marsupial species best studied, i.e., the North American opossum, *Didelphis virginiana*, and the Brazilian gray short-tailed opossum, *Monodelphis domestica*, long propriospinal projections (Cassidy and Cabana 1993), ascending spinal projections (Martin et al. 1983; Qin et al. 1993; Wang et al. 1997) and descending supraspinal pathways (see Sect. 7.3) as well as simple sensorimotor reflexes (Cassidy et al. 1994) and quadrupedal locomotion (Pflieger et al. 1996), largely form postnatally. *Monodelphis* has a longer gestational period than *Didelphis*, and its brain is more mature at birth. In *D. virginiana*, spinal axons innervate the medullary reticular formation by P7, but do not reach the diencephalon until P30 (Martin et al. 1983). Spinal axons reach the dorsal column nuclei by P5 (Wang et al. 1997), whereas the dorsal column nuclei innervate the ventral thalamus by at least P17 (Martin et al. 1987). Thalamocortical axons do not grow into the presumptive somatosensory/motor cortex until P21. Cortical axons do not innervate the spinal cord before P30. In *Monodelphis*, spinal projections reach the brain stem a few days earlier (Qin et al. 1993), and thalamocortical projections reach the various parts of the cerebral cortex between P5 and P9 (Molnár et al. 1998). About P10, the major corticofugal projections (corticotectal, corticopontine, and corticospinal axons) have extended through the internal capsule and form the cerebral peduncle (Molnár et al. 1998), whereas corticospinal fibers reach the cervical cord at P17. In *Monodelphis*, the period P0-P15 corresponds to that of E12-P0 in rats (Molnár et al. 1998).

7.2
Birthday Studies

Neuronal birthday data are sparse in opossums. Unpublished observations in the North American opossum by Johnson, Cabana and Martin (quoted from Cabana and Martin 1984) suggest that the reticular formation, the raphe nuclei and the locus coeruleus are among the first to enter the postmitotic period with the red nucleus clearly lagging behind. The brainstem raphe and adjacent part of the reticular formation contain serotonergic neurons in the newborn (Humbertson and Martin 1979; Humbertson et al. 1982; Martin et al. 1991a), and catecholaminergic neurons are already present at birth in the locus coeruleus (Martin et al. 1978; Pindzola et al. 1990). Saunders et al. (1989) studied the development of the neocortex in *Monodelphis domestica*. The cortical plate does not begin to appear until P3–P5.

7.3
Tract-Tracing Data in Opossums

Tract-tracing studies on the development of descending supraspinal projections were carried out especially in the North American opossum by Martin and co-workers, but more recently also in the more easily bred Brazilian gray short-tailed opossum. The development of descending supraspinal projections in these two opossum species is summarized in Table 6.

Table 6. Development of descending supraspinal projections in opossums

Nuclei	First appearance of monoaminergic neurons	Innervation of lumbar cord in Didelphis virginiana[e]	Innervation of lumbar cord in Monodelphis domestica[g]
Reticular formation			
Medullary		P1-P3	P1
Pontine		P1	P1
Interstitial nucleus flm		P1-P3	P1
Raphe nuclei	E11[c]	P1	P1
Serotonergic projections[d]		P1	
Vestibular nuclei			
Lateral vestibular nucleus		P1–P3	P1
Medial and inferior nuclei		P3–P5	P4
Locus coeruleus			
Coeruleospinal neurons	Prenatally[f]	P1	P1
Noradrenergic projections[f]		P1	
Red nucleus		P10[b]	P4
Hypothalamus			
Paraventricular nucleus		P3	P3
Corticospinal projections		P30[a] (cervical cord)	P17 (cervical cord)
Time of birth		12–13 days after conception	14–15 days after conception

[a] Cabana and Martin 1985.
[b] Cabana and Martin 1986a.
[c] Humbertson et al. 1982.
[d] Martin et al. 1991a.
[e] Martin et al. 1993.
[f] Pindzola et al. 1990.
[g] Wang et al. 1992.

7.3.1
Retrograde Tracer Data

Using the retrograde transport of HRP and the fluorescent tracer Nuclear Yellow, Cabana and Martin (1982, 1984) showed that in the North American opossum neurons within the pontine reticular formation and the presumptive locus coeruleus project to the thoracic spinal cord by P3. The application of another, more powerful fluorescent marker, Fast Blue, showed that already at P1 a number of brainstem structures and even the hypothalamus innervate the lumbar spinal cord (Martin et al. 1991b, 1993). At P1, supraspinal labeling was relatively sparse, and restricted to the ventral part of the rostral hypothalamus, the reticular formation, the coeruleus com-

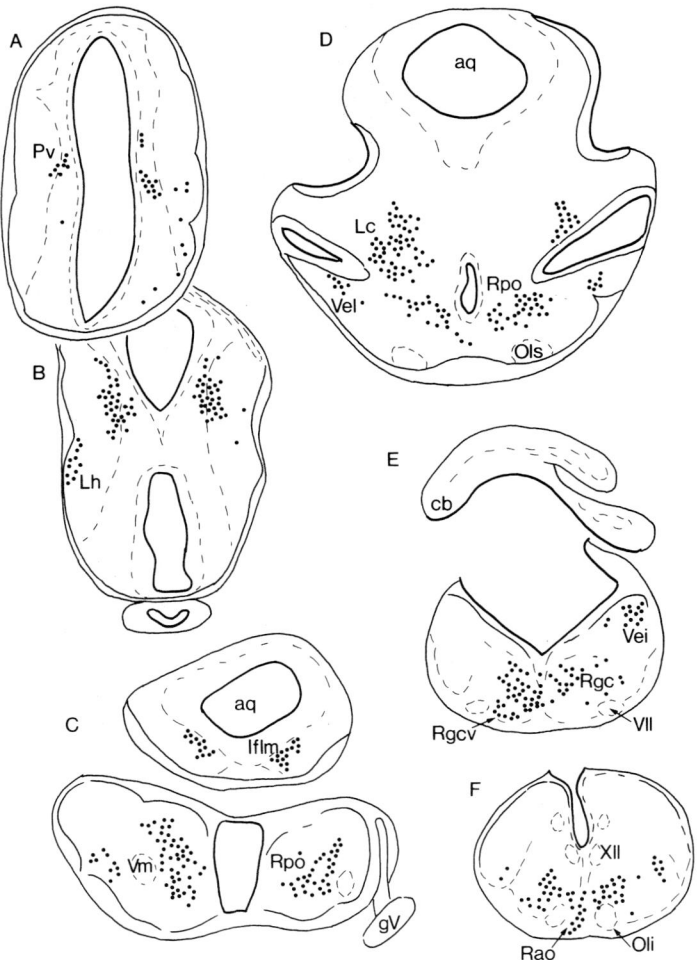

Fig. 32. The distribution of retrogradely labeled neurons in the hypothalamus and brain stem of the North American opossum, *Didelphis virginiana*, after a Fast Blue injection into the lumbar cord at P3. *aq*, aqueduct; *cb*, cerebellum; *gV*, ganglion of trigeminal nerve; *Iflm*, interstitial nucleus of flm; *Lc*, locus coeruleus; *Lh*, lateral hypothalamus; *Oli*, inferior olive; *Ols*, superior olive; *Pv*, paraventricular nucleus; *Rao*, nucleus raphes obscurus; *Rgc*, *Rgcv*, nucleus reticularis gigantocellularis and its ventral part, respectively; *Rpo*, nucleus reticularis pontis; *Vei*, *Vel*, inferior and lateral vestibular nuclei, respectively; *Vm*, *VII*, *XII*, motor nuclei of fifth, seventh and twelfth cranial nerves, respectively. (After Martin et al. 1993)

plex, and the caudal raphe. The ventral hypothalamic labeling was not observed in later stages of development and may represent a transient projection. At P1, the heaviest and most consistent labeling was found in the pontine reticular formation and the coeruleus complex. This rather heavy labeling suggests that reticulospinal projections are already present before birth. In one case, Martin et al. (1993) observed additional labeling in the interstitial nucleus of the flm and the lateral vestibular nucleus. Only a few propriospinal neurons were labeled rostral to the injection site. At P3, supraspinal and propriospinal labeling was much more abundant (Fig. 32). In the hypothalamus, labeling was found in the paraventricular nucleus and the lateral hypothalamic area. In the mesencephalon, labeled neurons were present in the interstitial nucleus of the flm, in the pons they were abundant in the reticular formation, the coeruleus complex, and the lateral vestibular nucleus. More caudally, descending supraspinal projections were found to arise from the inferior vestibular nucleus, the gigantocellular and ventral gigantocellular reticular nuclei, the central reticular nucleus, the retroambiguus nucleus, and from the raphe magnus, pallidus and obscurus. At P5, additional labeling was noted in the medial vestibular nucleus, the solitary complex, and deep to the dorsal column nuclei. By P10, labeled neurons were also found in the red nucleus (see also Cabana and Martin 1986a). Cortical neurons were not labeled from the cervical cord until about 30 days after birth (Cabana and Martin 1984, 1985, 1986b).

In the gray short-tailed opossum, at P1 several areas of the reticular formation, the caudal raphe, the coeruleus complex, and the lateral vestibular nucleus innervate the lumbar spinal cord (Wang et al. 1992). In the reticular formation, the interstitial nucleus of the flm, the pontine reticular nucleus, the gigantocellular and lateral paragigantocellular reticular nuclei, the ventral and dorsal medullary reticular nuclei and the retroambiguus nucleus were found labeled. Projections from the medial and inferior vestibular nuclei as well as the red nucleus reach the lumbar spinal cord by P4, and those from the paraventricular nucleus in the hypothalamus one day earlier. Cortical labeling after cervical tracer injections was not present before P17. Projections from the lateral, medial and inferior vestibular nuclei to the cervical spinal cord are present at birth (Pflieger and Cabana 1996). These data show that in opossums: (1) restricted areas of the brain stem innervate the lumbar spinal cord at birth, and must have reached the cervical spinal cord prior to birth; (2) several brainstem nuclei do not project to the spinal cord until well after birth; and (3) that the development of descending supraspinal pathways occurs asynchronously and according to a predictable sequence (Martin et al. 1991b; Wang et al. 1992).

7.3.2
The Ingrowth of Reticulospinal, Monoaminergic and Rubrospinal Fibers

Degeneration studies (Martin et al. 1978) revealed that immature brainstem axons are present within the marginal zone of the lumbosacral spinal cord of the North American opossum before hindlimb movements begin (behavioral stage I). Some of these fibers are monoaminergic, and presumably arise in the locus coeruleus. Immature synapses were found in the marginal zone. By the time random hindlimb movements can be observed (stage II), brainstem axons including monoaminergic fibers had grown into limited areas of the intermediate zone of the lumbosacral spinal cord. Such axons are still immature, but it remains unclear that brainstem axons contribute to the synapses present in both the marginal and intermediate zones at this stage of development. Brainstem axons continue to grow into the intermediate zone of the lumbosacral spinal cord and occupy all of their adult territories before thoracic transection produces obvious change in hindlimb motility. It takes another 20 days before thoracic transection produces spinal shock comparable to that in the adult animal (see Martin et al. 1975).

In *Monodelphis domestica*, Cabana and co-workers studied the synaptogenesis in the spinal cord using three synaptic markers (synaptophysin, synaptotagmin-1, and SNAP-25). Synaptogenesis in the lumbosacral spinal cord correlates with the arrival of the major descending pathways, and is particularly intense in the ventral horn when the hindlimbs begin to move, and in the dorsal horn when simple hindlimb reflexes such as grasping and withdrawal can be elicited (Cabana et al. 1998). In the brachial and lumbosacral enlargements, the three synaptic markers were first observed in the white matter, presumably in growing axons, and then in the gray matter according to rostrocaudal and ventrodorsal gradients (Gingras and Cabana 1998).

In *Didelphis virginiana*, Martin and co-workers extensively studied the development of monoaminergic projections to the spinal cord. In newborn opossums, serotonergic axons were found at most spinal levels (Humbertson and Martin 1979; Humbertson et al. 1982; DiTirro et al. 1983). Such axons were present predominantly in the marginal zone, but a few were found in the intermediate zone at cervical levels. By P5, serotonergic axons were present in most areas of the intermediate zone, particularly at cervical and thoracic levels, except for the superficial layers I and II of the dorsal horn. By P15, serotonergic axons were abundant within layers IV–X as well as

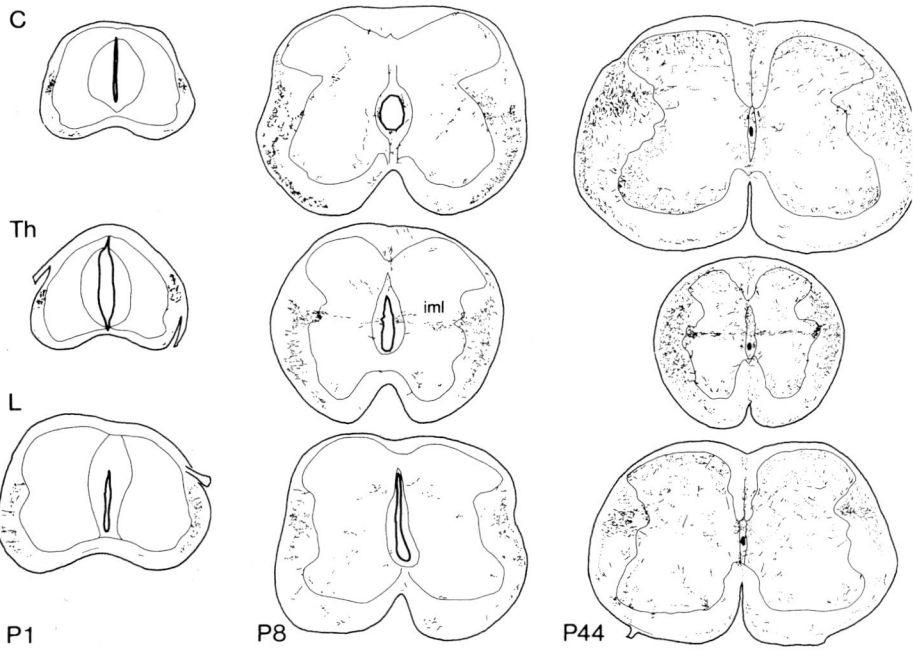

Fig. 33. The distribution of tyrosine hydroxylase-immunoreactive axons in the spinal cord of the North American opossum at P1, P8, and P44. *C, L, Th*, cervical, lumbar and thoracic spinal segments, respectively; *iml*, intermediolateral nucleus. (After Pindzola et al. 1990).

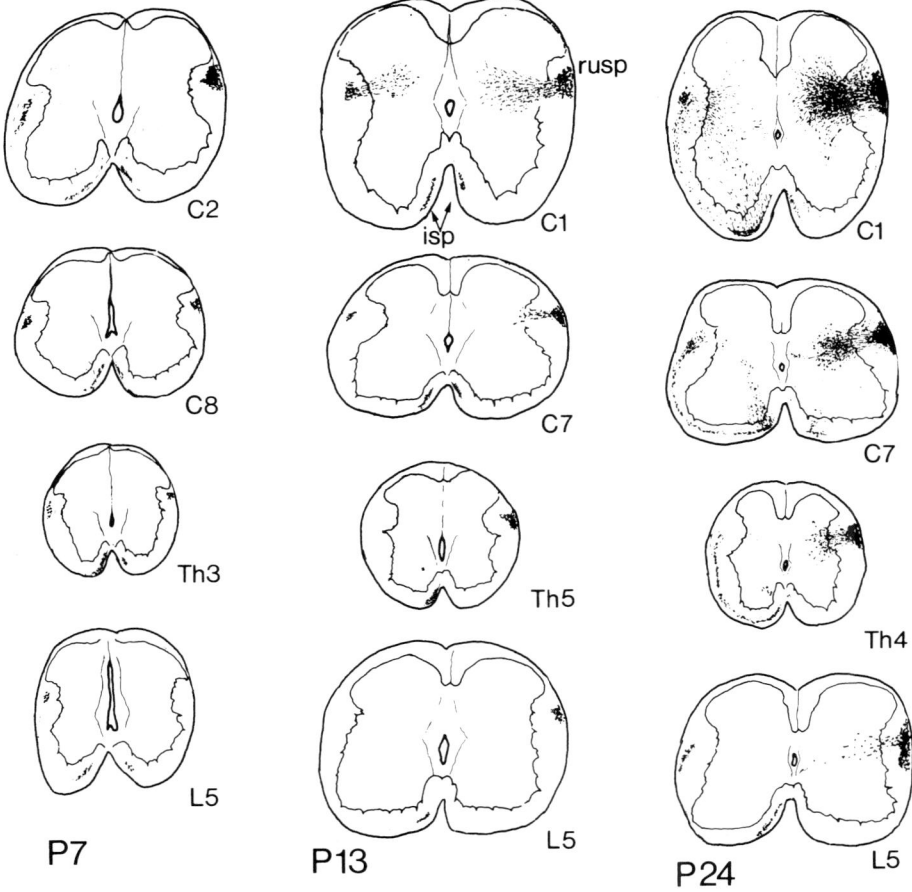

Fig. 34. The development of the rubrospinal tract in the North American opossum. *rusp*, rubrospinal tract. (After Cabana and Martin 1984, 1986a, 1988)

in the motoneuronal layer IX. The superficial layers I and II were not innervated before P50. At that time the intraspinal distribution of serotonergic axons was comparable to that in adult opossums (see Martin et al. 1982b).

Using an antibody against tyrosine hydroxylase, Pindzola et al. (1990) demonstrated that catecholaminergic projections are present in the spinal cord of newborn opossums (Fig. 33). At birth, such axons are particularly abundant in the dorsolateral marginal zone, the region containing most of them in adult animals (see Pindzola et al. 1988). At P3, a few immunostained fibers were found in the intermediate zone of the spinal cord, and at P8, they were concentrated in the presumptive intermediolateral cell column. The superficial layers I and II are not innervated until about P15. By P44, the distribution of tyrosine hydroxylase-immunoreactive fibers in the spinal cord resembles that in adult animals.

In opossums, the development of the rubrospinal tract occurs postnatally (Cabana and Martin 1982, 1984, 1986a; Wang et al. 1992; Martin et al. 1993). In the North American opossum, Cabana and Martin (1986a) observed the first rubrospinal axons in the cervical spinal cord around P5 (Fig. 34), and in the thoracic cord by P7, in line with retrograde tracing experiments (Cabana and Martin 1984). The adult pattern of rubrospinal innervation is reached by P35. Rubrospinal axons do not grow synchronously as a massive bundle following a few leading fibers, but by gradual addition of axons. Rubrospinal axons influence the spinal cord early in development, much earlier than cortical axons (see Sect. 7.3.4), and are already under cerebellar control by P13 (Martin et al. 1986, 1987, 1988).

7.3.3
Regenerative Capacity of Descending Brainstem Pathways

Opossums have been used to study the plasticity of descending brainstem pathways (Martin and Xu 1988; Xu and Martin 1989, 1991, 1992; Treherne et al. 1992; Woodward et al. 1993; Martin et al. 1994; Nicholls et al. 1994; Wang et al. 1994, 1996; Varga et al. 1995a,b; Nicholls and Saunders 1996; Saunders 1997; Saunders et al. 1998). In newborn opossums, repair of spinal cord connections occurs rapidly. As the CNS matures, the capacity for regeneration ceases abruptly. In *Didelphis virginiana*, initial studies focussed on the rubrospinal tract (Martin and Xu 1988; Xu and Martin 1989, 1991, 1992). Spinal cord lesions at P18 or later were combined with retrograde and anterograde tracing experiments. It appeared that only a few rubrospinal neurons survived axotomy, suggesting that the growth of new, later arising axons around the spinal lesion is a major factor in developmental plasticity. Ghooray and Martin (1993a) found a rough temporal correlation between the development of myelin, as demonstrated by the presence of myelin basic protein, and the end of the critical period for rubrospinal plasticity. Moreover, a temporal correlation exists between the transition from immature to mature appearing glia and loss of rubrospinal plasticity (Ghooray and Martin 1993b). Myelin basic protein (MBP) immunoreactivity was first observed in the opossum's brain stem (in the flm) and spinal cord at P15, and its development in most tracts followed rostral to caudal gradients. At P15, it was present in the cervical and upper thoracic spinal cord in the ventral funiculus, the location of pontine reticulospinal and lateral vestibulospinal axons, at P21 throughout the length of the spinal cord in the ventral funiculus, and at cervical levels also in the ventral part of the lateral funiculus and cuneate fascicle. At P26, at cervical levels sparse MBP-immunoreactivity was found in the dorsal part of the lateral funiculus, the area which contains rubrospinal axons. By P30, MBP-immunoreactivity was found in the rubrospinal tract at thoracic levels, at P33 also at lumbosacral levels. The fasciculus gracilis was labeled at lumbar levels at P30, and at thoracic levels at P33. The dorsal corticospinal tract was immunoreactive by P54. MBP-immunoreactivity was present well before myelin can be identified with the Weigert technique (Langworthy 1929). The critical period for rubrospinal plasticity usually ends between P26 and P30 (Xu and Martin 1989, 1991, 1992). Reticulospinal and vestibulospinal axons, like rubrospinal axons, grew around a lesion of their pathway during development (Wang et al. 1994). The critical period for their plasticity ends earlier than that for rubrospinal axons, usually sometime between P12 and P20 (Wang et al. 1994). These data are consistent with the hypothesis that axons reaching the spinal cord early, lose their potential for plasticity before axons arriving later. More recent studies (Wang et al. 1996, 1998a) showed that, although late growth appears to be a major contributor to such plasticity, true regeneration of cut axons also occurs. Supraspinal axons grow across the lesion after complete transection of the thoracic spinal cord when the lesion is made at younger ages than those used in previous experiments (Wang et al. 1996). The critical period for such plasticity also differs for different pathways (Wang et al. 1996). Regeneration of cut axons contributes to growth across the lesion (Wang et al. 1998a). Axons which grow across the lesion innervate areas that are appropriate for them and they contribute to the relatively normal use of the hindlimbs at maturity (Wang et al. 1998b). Adult opossums which had their thoracic spinal cord transected on P5 exhibit remarkably normal use of the hindlimbs in locomotion as well as sensation caudal to the lesion. Growth of descending supraspinal and propriospinal axons through a spinal cord lesion occurs when a tissue bridge develops across the lesion site.

Ascending spinal axons are also capable of developmental plasticity (Terman et al. 1996, 1997, 1999; Wang et al. 1997). Axons of all ascending spinal pathways grow through a thoracic spinal cord lesion, and reach appropriate supraspinal areas, if the lesion is made early enough in development. The critical period for plasticity of ascending axons ends before P12, i.e., well before that of most descending spinal axons. This indicates that inhibitory factors in myelin are not the only ones which impede developmental plasticity (Terman et al. 1999).

In *Monodelphis domestica*, an isolated CNS preparation was used for determining the critical period for repair of connections in injured neonates (Treherne et al. 1992; Woodward et al. 1993; Varga et al. 1995a, b). In postnatal opossums (P3-P6), repair was observed 5 days after lesions were made in vitro at the cervical level. Through-conduction of action potentials was re-established and axons stained by DiI grew into and beyond the lesion. At P11–12, repair was more limited, whereas at P13–14 none of the preparations examined showed any axonal growth into the lesion or conduction through it (Varga et al. 1995a). The critical period at which regeneration stops coincides with myelination and oligodendrocyte development. This period can be extended by blocking myelin-associated inhibitors (Varga et al. 1995b). In vitro studies suggest that the onset of expression of myelin-associated neurite growth inhibitors occurs as late as P12–P14 in the lower cervical spinal cord of *Monodelphis* (Varga et al. 1995a,b). Saunders et al. (1998) showed that complete transection of the thoracic spinal cord in the first postnatal week of *Monodelphis* was followed by fiber growth across the lesion, with substantially normal development of spinal cord structure, impulse conduction, and locomotor behavior. After a spinal injection of a dextran amine (RDA) caudal to the lesion retrogradely labeled neurons were found in many nuclei shown previously to project to the spinal cord (Holst et al. 1991; Wang et al. 1992), including the reticular formation, the lateral vestibular nucleus, the locus coeruleus, raphe nuclei, the red nucleus, and the interstitial nucleus of the flm.

7.3.4
Development of Corticospinal Projections in Opossums

In the North American opossum, corticospinal axons only project to cervical and rostral thoracic levels of the spinal cord (Martin and Fisher 1968). Martin and co-workers (Martin et al. 1980; Cabana and Martin 1985, 1986a,b) showed that cortical axons reach the ventral mesencephalon by P10, and start decussating in the caudal medulla by P23. Corticospinal axons first enter the spinal cord at P28, and are present in the white matter before growing into the dorsal horn. Corticospinal axons do not grow into the gray matter until several days later and do not reach their caudal extent (Th4–5) until P38. At about P30, retrogradely labeled corticospinal neurons (Fig. 35) were limited to a small area caudal to the orbital sulcus corresponding to part of the primary somatosensory/motor cortex (see Lende 1963a,b). During later stages, neurons were labeled over a wider area of cortex than in adult opossums. Corticospinal axons grow into the gray matter exclusively from the dorsal and lateral funiculi and first innervate adjacent portions of laminae IV and V. They subsequently extend into laminae III and VI, then VII, VIII and X and, finally, in the cervical enlargement, the medial edge of laminae I–II as well as lamina IX. There is a subsequent period of development during which the density of cortical innervation in all spinal laminae, particularly in the ventral horn, appears to exceed that in the adult opossum (Cabana and Martin 1985). *Monodelphis domestica* has an extremely modest corticospinal projection (Nudo and Masterton 1990a,b; Holst et al. 1991). After cervical Fast Blue injections cortical labeling was present at P17 (Wang et al. 1992).

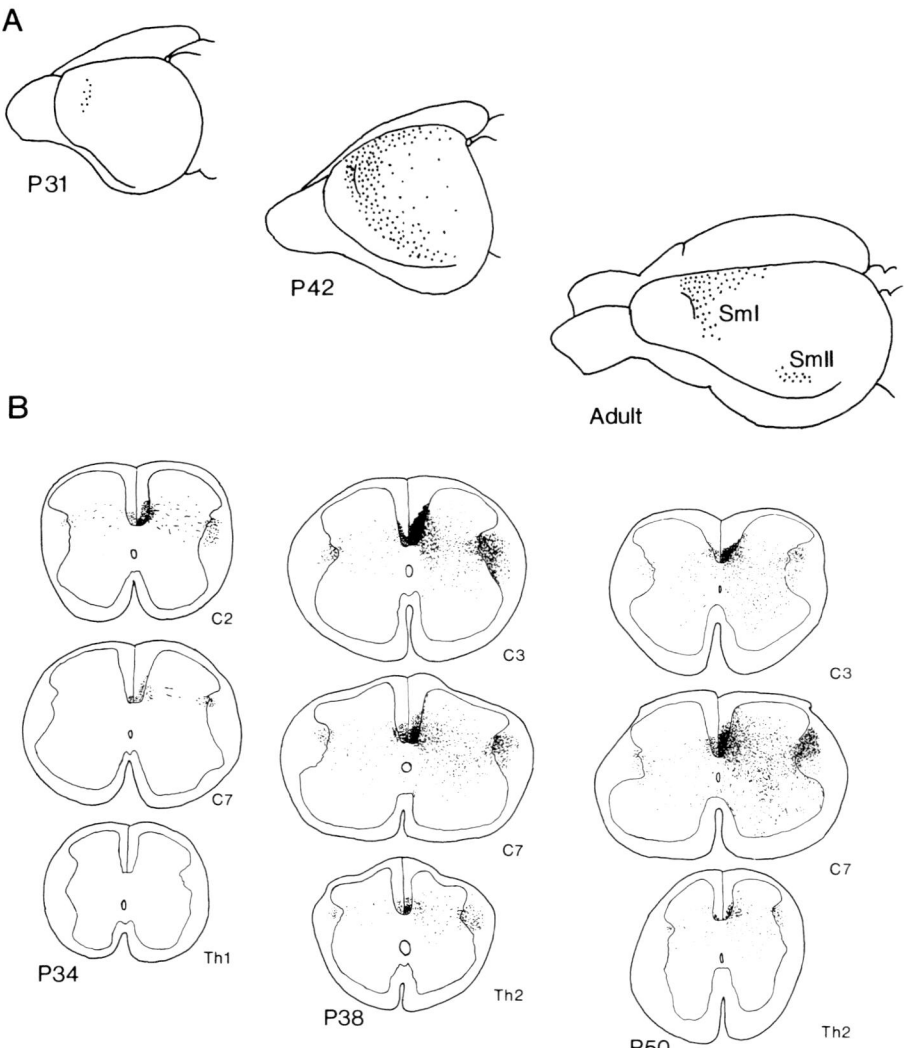

Fig. 35A,B. The development of the corticospinal tract in the North American opossum. In **A** the distribution of the cells of origin is shown for P31, P42, and an adult. In **B** the distribution of corticospinal fibers and their terminals is shown for P34, P38, and P50. *SmI, SmII*, primary and secondary somatomotor cortex, respectively. (After Cabana and Martin 1984, 1985)

8 Development of Descending Supraspinal Pathways in Placental Animals

8.1
Initial Tract Formation

Windle's studies on the development of neurofibrils in axonal tracts in mammals showed that the fasciculus longitudinalis medialis (flm) is one of the earliest differentiating fiber tracts (Windle 1932a,b, 1935; Windle and Baxter 1936; Rhines and Windle 1941). In pyridine silver stained rat, cat and human embryos, the first, rostral component of the flm arises from a nucleus at the mesodiencephalic junction at E11 (approximately 272 h after insemination) in the rat, at about E16 in the cat, and between E26 and E30 in man. In rat embryos, Bélanger et al. (1993) observed two longitudinal columns of early-generated brainstem neurons. These longitudinal columns were associated with well-differentiated parts of the marginal zone representing the prospective sites of the flm and the lateral longitudinal tract. Axons were seen to travel in the early flm and the lateral longitudinal tract in close proximity to the early-generated neurons. These pathways are reminiscent of the basal and alar substrate pathways found by Katz et al. (1980) in anurans. With an antibody to neuron-specific class III β-tubulin, Easter et al. (1993) showed that in the mouse the flm arises at E9.5 (Fig. 36). The first labeled neurons were found in the basal plate near the cephalic flexure. Caudally directed axons emerge from this group to form the flm. In mice, in contrast to other vertebrates (Easter et al. 1994), the first tract to develop is the descending tract of the mesencephalic trigeminal nucleus, found to arise from the alar plate at E8.5. In the most advanced E9.5-embryos, the flm was found to extend to the level of entrance of the trigeminal nerve, and by E10, beyond the level of the otic placode. It should be noted that the development of the mouse brain precedes that of the rat by about two days (see Table 2).

8.2
Tract-Tracing Data in Rodents

In rodents, most descending supraspinal pathways arise prenatally. The major ascending spinal pathways are also well-established by the day of birth. In striking contrast, the corticospinal tract does not enter the spinal cord before birth, and its outgrowth throughout the cord occurs entirely postnatally. The development of the corticospinal tract appears to correlate with the appearance of forelimb and hindlimb placing

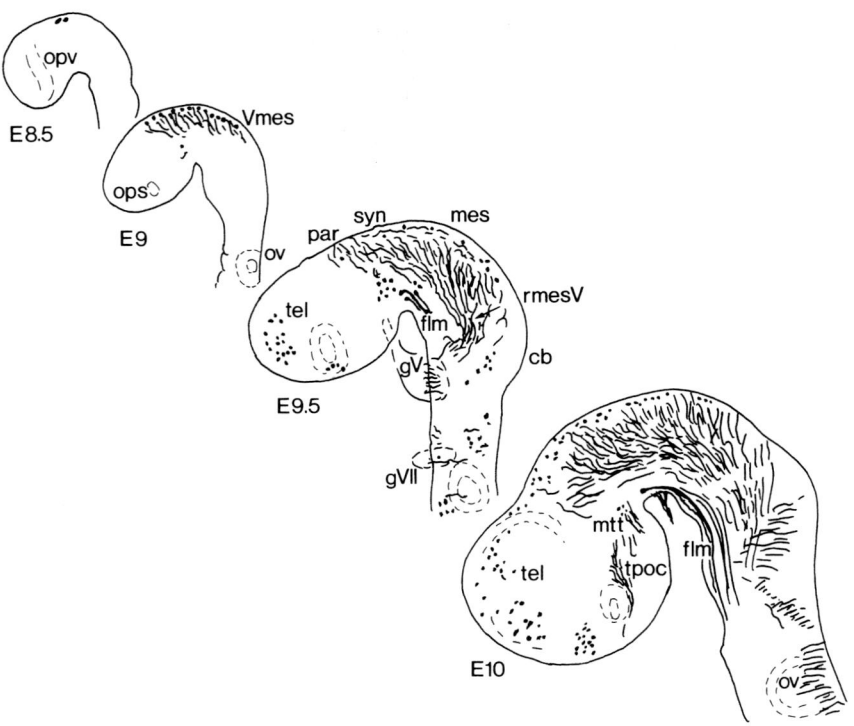

Fig. 36. Initial tract formation in the mouse brain. *cb*, cerebellum; *flm*, fasciculus longitudinalis medialis; *gV*, *gVIII*, ganglia of Vth and VIIIth nerves, respectively; *mes*, mesencephalon; *mtt*, mamillotegmental tract; *ops*, *opv*, optic stalk and vesicle, respectively; *ov*, otic vesicle; *par*, parencephalon; *rmesV*, mesencephalic root of Vth nerve; *syn*, synencephalon; *tel*, telencephalon; *tpoc*, tract of postoptic commissure; *Vmes*, mesencephalic nucleus of Vth nerve. (After Easter et al. 1993)

reactions (Hicks and D'Amato 1975; Donatelle 1977). Moreover, in rats spontaneous quadrupedal walking was first observed at P11 (Westerga and Gramsbergen 1990).

In rats, spinal neurons are generated in a sequential order from ventral to dorsal (Altman and Bayer 1984): first, the basal plate generates the motoneurons, followed by the intermediate plate that generates the relay neurons in the intermediate zone, and finally the alar plate produces the interneurons in the dorsal horn. Motoneurons are produced over a two-day period: peak production is on E12 at cervical levels, and on E14 at thoracic and lumbar levels. The bulk of dorsal root ganglion (DRG) cells is produced between E12 and E15 (Fig. 37). Large ganglion cells are generated before small ones. The earliest dorsal root fibers enter the spinal cord at E13 at cervical levels (Altman and Bayer 1984). With carbocyanine dyes, Snider et al. (1992) traced the outgrowth of dendrites of cervical motoneuron pools and the development of dorsal root projections to these motor pools (see also Mirnics and Koerber 1995). At E15, the first day at which dorsal root fibers could be seen entering the cervical cord, the lateral motoneurons extended their dendrites medially or dorsomedially into the direction of the incoming dorsal root fibers. Between E15 and E17 fascicles of dorsal root axons

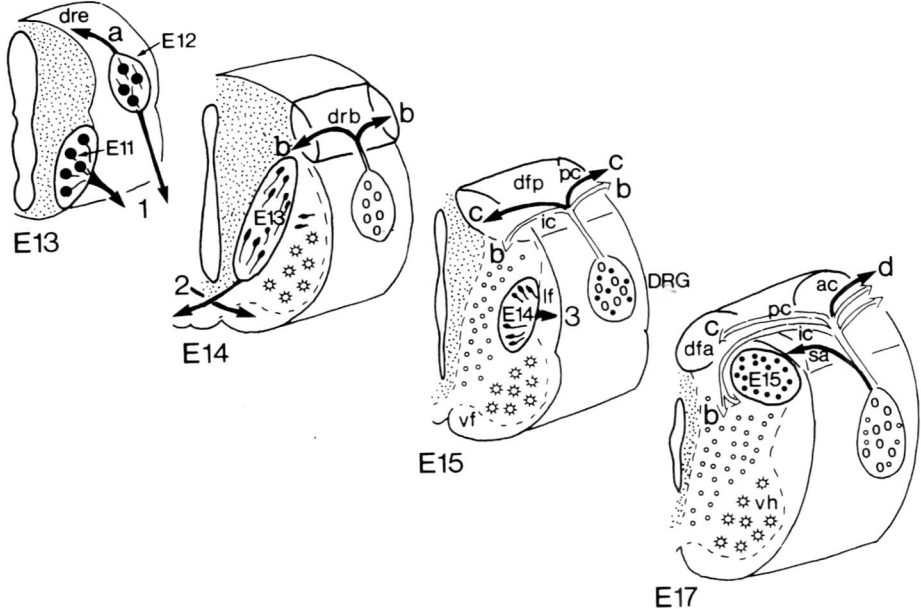

Fig. 37. Main developmental events (*heavy lines*) in the rat cervical spinal cord from E13 until E17. At *E13*, onset of growth of peripheral motor fibers from the early generated motoneurons (*E11*), and of sensory fibers from the early generated dorsal root ganglion cells (*E12*) takes place (*1*). At E14, the ventral commissure and ventral funiculi are formed (*2*) by the axons of contralaterally projecting relay neurons that are generated on E13. At *E15*, lateral migration of the ipsilaterally projecting relay neurons (generated predominantly on *E14*) occurs (*3*). Interneurons in the dorsal horn and small dorsal root ganglion cells are generated on E15. At *E17*, the dorsal funicular ascending zone (*dfa*) is formed. *a–d*, the ingrowth of dorsal root fibers: *a* arrival of the earliest dorsal root fibers at the dorsal root entrance zone (*dre*), *b* formation of the dorsal root bifurcation zone (*drb*), *c* formation of the dorsal funiculus propriospinal zone (*dfp*) by the growing intersegmental dorsal root collaterals, and *d* formation of the dorsal funiculus ascending zone (*dfa*) by the growing suprasegmental dorsal root collaterals. *ac*, ascending dorsal root collaterals; *ic*, intrasegmental collaterals; *lf*, lateral funiculus; *pc*, propriospinal (intersegmental) collaterals; *sa*, small-caliber collaterals, *vf*, ventral funiculus; *vh*, ventral horn. (After Altman and Bayer 1984)

converge in the intermediate zone and fan out en route to the motor pools. Between E17 and E19 there is dense branching and bouton formation of muscle (Ia) afferents in the area of the motor pools (Snider et al. 1992).

Several types of interneurons can be distinguished in the rat spinal cord (Silos-Santiago and Snider 1992, 1994; for mouse data see Wentworth 1984b). In the thoracic spinal cord, Silos-Santiago and Snider (1992) noted seven different types of commissural interneurons, i.e., interneurons with a contralaterally projecting axon, by E13.5. By E15, commissural interneurons were found near their final locations in the dorsal horn, the intermediate zone and the ventral horn. By E19, at least 18 different types of commissural interneurons were found. Also an increasing number of ipsilaterally projecting interneurons were found in the thoracic spinal cord from E14 till E19

81

(Silos-Santiago and Snider 1994). Therefore, the rat embryonic spinal cord contains a large number of ipsilaterally projecting as well as commissural interneurons.

In mice, motoneurons are generated on E10 and E11, neurons in the intermediate zone from E11 through E14, and neurons in the dorsal horn from E12 through E14 (Nornes and Carry 1978). Large DRG cells are generated in peak numbers on E10.5, whereas small DRG cells arise in greatest numbers on E12 (Sims and Vaughn 1979). The first, axodendritic synaptic contacts on mouse lateral motoneurons, presumably propriospinal in origin, were found at E11 (Vaughn et al. 1977). Both axodendritic and axosomatic synapses were found at E12. Most of the early forming, lateral motoneuronal dendrites grow into the lateral marginal zone (see also Wentworth 1984a), where they come into contact with axons of interneurons. Ozaki and Snider (1997) studied the initial trajectories of sensory axons to the thoracic spinal cord. Primary afferent axons reach the thoracic cord at E10.5, and grow rostrocaudally for at least 48 h prior to extending collateral branches into the gray matter. Such a "waiting period" has been documented also in anurans (Smith and Frank 1988a) and chickens (Davis et al. 1989b). The onset of collateral branching does not occur until well after differentiation of certain cell groups in the developing dorsal part of the cord (Ozaki and Snider 1997). Different classes of murine primary afferent fibers enter the spinal cord in sequence (Ozaki and Snider 1997): muscle afferents penetrate the gray matter as early as E13.5, large-caliber sensory afferents first penetrate at E14.5, and most fine cutaneous afferents enter at E15.5.

Ascending spinal tract neurons begin to differentiate as early as E12 (Altman and Bayer 1984). Waldeyer's cells in the marginal zone, a major source of contralaterally projecting spinothalamic fibers, and several other cells in the intermediate zone giving rise to spinocervical, rostral and ventral spinocerebellar, and some spinothalamic fibers are produced on E12 and E13. The neurons of Clarke's column giving rise to the dorsal spinocerebellar tract are also formed on E13. Most neurons of the intermediate zone are generated on E13 and E14. Beal and Bice (1994) showed that lumbar spinothalamic and spinocerebellar neurons are generated between E13 and E15. The primary afferent projections from dorsal root fibers to the dorsal column nuclei also arise prenatally (Chimelli and Scaravilli 1987; Wessels et al. 1991). Projections from the dorsal column nuclei reach the thalamus by the day of birth (Asanuma et al. 1988). Lakke (1997) found the first spinothalamic fibers, labeled from cervical injections, in the anterior thalamus at E18.

Thus, in rats the earliest propriospinal and ascending projections appear to be formed by E12, i.e., around the same time that motoneurons are developing and descending supraspinal projections start to invade the spinal cord. Dorsal root projections do not enter the spinal gray matter before E15.

In rats (see Table 7), early brainstem-spinal cord projections were studied using the carbocyanine dye DiI in fixed embryos (Auclair et al. 1993, 1999; de Boer-van Huizen and ten Donkelaar 1999), and BDA in an isolated embryonic brain-spinal cord preparation (de Boer-van Huizen and ten Donkelaar 1999). With both techniques it was shown that in embryos at least 12 days of age (E12), i.e., at the time of closure of the posterior neuropore (stage 12, see Table 2), a variety of brainstem centers already innervates the spinal cord. In the interstitial nucleus of the flm and various parts of the reticular formation – mesencephalic, pontine as well as medullary – DiI or BDA labeled neurons were observed (Fig. 38). Mainly large immature, bipolar neurons were labeled. In later stages (E13, E14) the number of labeled neurons increased and

Table 7. The development of descending supraspinal projections in rats

Nuclei	Time of neuron origin[a]	Innervation of spinal cord			
		High cervical[b]	Midcervical[c]	Lower thoracic cord[d]	Lumbosacral cord[e]
Reticular formation					
Medullary	E11–E15	E12	E14	E15	E17/19
Pontine	< E11–E15	E12	E13	E14	E18
Mesencephalic	?	E12	?	E14	E18
Interstitial nucleus flm	?	E12	E13	E14	E18
Raphe nuclei	E11–E15	E14	E14	E15	E17
Serotonergic projections[f]		E14	E14	E15	E17
Vestibular nuclei					
Lateral vestibular nucleus	E11–E14	E13	E13	E14	E17
Medial and inferior vestibular nuclei	E12–E15	?	?	E20	E18–E21
Locus coeruleus					
Coeruleospinal neurons	peak E12	E12?	E13?	E15	E20
Noradrenergic projections[g]		< E16	< E16	E16/17	E17/18
Red nucleus	E13–E14	E17[h]	E18[h]	E19[h]	E21[h]
Hypothalamus					
Paraventricular nucleus	Peak E14–15			E20.5	P1
Corticospinal projections[i]	Peak E15–17	P0	P1	P5	P7–P9

[a] Altman and Bayer 1980a–d, 1981, 1995, Bayer and Altman 1995.
[b] de Boer-van Huizen and ten Donkelaar 1999.
[c] Auclair et al. 1993, 1999.
[d] Kudo et al. 1993.
[e] Lakke 1997.
[f] Rajaofetra et al. 1989.
[g] Rajaofetra et al. 1992.
[h] Lakke and Marani 1991.
[i] Gribnau et al. 1986.

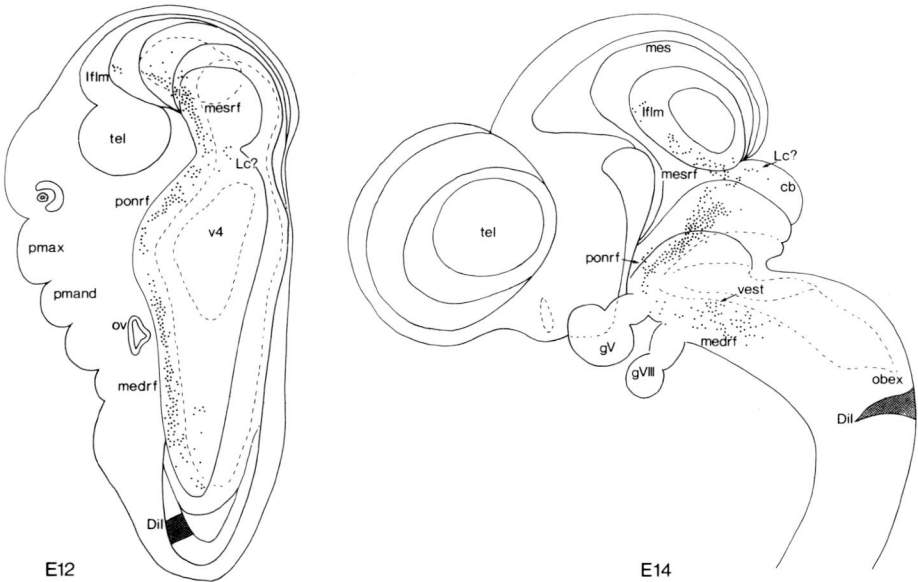

Fig. 38. The distribution of labeled brainstem neurons in an E12 and an E14 rat embryo following the application of a DiI crystal just behind the obex shown in reconstructions of a nearly horizontally sectioned (*E12*) and a sagittally sectioned (*E14*) brain. *cb*, cerebellum; *gV*, *gVIII*, trigeminal and vestibulocochlear ganglia; *Iflm*, interstitial nucleus of flm; *Lc?*, locus coeruleus?; *medrf*, *mesrf*, *ponrf*, medullary, mesencephalic and pontine parts of reticular formation, respectively; *mes*, mesencephalon; *ov*, otic vesicle; *pmand*, *pmax*, mandibular and maxillary processes, respectively; *tel*, telencephalon; *vest*, vestibular nuclear complex; *v4*, fourth ventricle. (After de Boer-van Huizen and ten Donkelaar 1999)

more mature, multipolar cells were found. At E13 (stage 15), labeled neurons were also observed in the vestibular nuclear complex. Raphespinal neurons were not labeled before E14 (stage 17). Just below the cerebellum a conspicuous small group of neurons was found labeled in a position reminiscent of the locus coeruleus. Birthday studies (see Table 7) in rats (Altman and Bayer 1980a–d, 1981, 1995; Bayer and Altman 1995) revealed that neurons in the medullary reticular formation are produced between E11 and E15 along a caudorostral gradient (Altman and Bayer 1980a,b), whereas those in the pontine reticular formation are generated even earlier (Altman and Bayer 1980d). In the vestibular nuclear complex, the large neurons in the lateral (Deiters) nucleus are generated before the smaller neurons in the other vestibular nuclei (Altman and Bayer 1980c). Neurons in the locus coeruleus are produced mostly on E12 (Altman and Bayer 1980d). No data are available on the interstitial nucleus of the flm, but in the related nucleus of Darkschewitsch peak production is on E12 and E13 (Altman and Bayer 1981). It should be emphasized that Altman and Bayer (1995) used another system of dating embryos than that employed in the present and other studies on the ingrowth of supraspinal fibers into the spinal cord. They considered the day when sperm were found in the vagina of a mated female as E1 and, consequently, E12 as embryonic day 12. In the tract-tracing studies (Auclair et al. 1993, 1999; de Boer-van Huizen and ten Donkelaar 1999), however, the day of mating was recorded as E0. This means that E12 corresponds to an embryo at least 12 days of age, i.e., starting or in its embryonic day 13. Table 7 includes Altman and Bayer's original data. Comparison of the data on the time of origin with those on the ingrowth of brainstem fibers into the cord (Table 7) suggests that interstitiospinal and reticulospinal neurons start projecting spinalwards shortly after they are generated. Since the distance to the site of tracer application for the interstitial nucleus of the flm and the pontine reticular formation by far exceeds that of the medullary reticular formation, it is most likely that interstitiospinal and pontine reticulospinal axons are the first supraspinal fibers to invade the spinal cord.

Kudo et al. (1993) studied the development of descending fibers to the lower thoracic spinal cord of rat embryos. Their data (see Table 7) show the gradual descent of supraspinal fibers into the spinal cord. The ingrowth of serotonergic and noradrenergic fibers in the rat spinal cord was studied with immunohistochemical techniques by Rajaofetra et al. (1989, 1992). The descent of serotonergic fibers is in line with the tract-tracing data available (Table 7). Noradrenergic projections to the spinal cord were studied from E16 onwards (Rajaofetra et al. 1992). The description of their data makes it likely, however, that noradrenergic fibers must have reached the cervical spinal cord before E16.

Lakke (1997) extensively studied the descent of descending supraspinal fibers in prenatal rats from E16 onwards (late embryonic and fetal rats) using an intrauterine approach. At E17, fibers from the lateral vestibular nucleus, the raphe magnus nucleus and the gigantocellular reticular nucleus had reached the lumbosacral spinal cord. At E18, among other nuclei, the interstitial nucleus of the flm, the mesencephalic reticular nucleus, the caudal pontine reticular nucleus, the subcoeruleus nucleus, the spinal vestibular nucleus, the nucleus raphes obscurus and the ventral medullary reticular nucleus had invaded the lumbosacral spinal cord. At E19, fibers from the oral pontine reticular nucleus, the ventral gigantocellular reticular nucleus and the nucleus ambiguus first appeared in the lumbosacral spinal cord. At E20, fibers from the locus coeruleus, the nucleus raphes pallidus and several reticular nuclei reached the lumbosacral spinal cord. Last to arrive prenatally (E21) in the lumbosacral spinal cord are axons from the red nucleus, the medial vestibular nucleus and the solitary nucleus. Fibers from the hypothalamus (the paraventricular nucleus and the lateral hypothalamic area) only arrived in the lumbosacral cord at P1. The prenatal descent of rubrospinal axons through the spinal cord of the rat was studied by Lakke and Marani (1991; see Table 7). Descending control at birth is exerted primarily on proximal muscles by ventral, reticulospinal pathways, whereas the control over distal muscles increases during the first days after birth (Brocard et al. 1997).

It is beyond the scope of this survey to discuss the vast literature on the development of the corticospinal tract in rodents (for reviews see Stanfield 1992; O'Leary and Koester 1993; Terashima 1995a; Joosten and Bär 1999). The corticospinal tract is the last of the descending supraspinal pathways to enter the spinal cord. In hamsters, corticospinal axons do not reach the spinal cord until several days after birth (Reh and Kalil 1981; Kuang and Kalil 1990), whereas in rats and mice corticospinal fibers reach upper cervical spinal segments at P0 (Donatelle 1977; Schreyer and Jones 1982; Terashima et al. 1983; de Kort et al. 1985; Gribnau et al. 1986; Joosten et al. 1989; Oudega et al. 1994). In rats, the postnatal development of the corticospinal tract can be divided into three periods: an outgrowth phase (P1–P10), a myelination phase (P10–P28), and a maturation phase (P28–adult). Gribnau et al. (1986) studied the outgrowth of corticospinal fibers in the rat spinal cord (Fig. 39). Rat corticospinal axons reach the third thoracic segment at P3, the upper lumbar cord at P7, and the sacral spinal cord at P9. After arrival of the first axons at a particular segment, new axons continue to be added to the tract for at least one week (Schreyer and Jones 1982, 1988; Gribnau et al. 1986; Gorgels et al. 1989). The timing of the outgrowth of the uncrossed corticospinal fibers in the ventral funiculus through the spinal cord appeared to be the same (Joosten et al. 1992). A delay of two days was found between the arrival of the corticospinal axons at a particular level of the spinal cord and their outgrowth into the spinal gray. Initially, most parts of the cortex including the occipital lobe innervate the spinal cord (Stanfield and O'Leary 1985; O'Leary and Stanfield 1986; Joosten et al. 1987; O'Leary and Koester 1993). Axons from frontal regions arrive first and those from the occipital lobe come last. Most of the transitory collaterals may not penetrate the spinal gray (Joosten et al. 1987). Curfs et al. (1994), however, demonstrated the presence of transient corticospinal projections to all parts of the cervical spinal gray matter between P4 and P10, before restriction to the adult pattern of termination. The withdrawal of collaterals probably accounts for the dramatic loss of fibers from the corticospinal tract during development (Reh and Kalil 1982a; Schreyer and Jones 1988). In the development of cortical axons arising in layer-V neurons three stages can be distinguished (O'Leary et al. 1990): (1) layer-V axons extend out of the cortex toward the spinal cord, bypassing their subcortical targets; (2) the subcortical targets are exclusively contacted by axon collaterals that develop by delayed interstitial branching off the flank of a spinally directed primary axon; and (3) specific branches and segments of the primary axon are selectively eliminated to yield the mature projections functionally appropriate for the area of cortex in question. In mutant rodents with extensive perturbations in the development of the cerebral cortex such as the reeler mouse and the shaking rat Kawasaki, corticospinal tract neurons are spread throughout all layers of the mutant cortex (Terashima et al. 1983; Inoue et al. 1991; Ikeda and Terashima 1997). The specificity of corticospinal connections is relatively unaffected (Terashima 1995a,b).

Direct corticomotoneuronal contacts in the cervical spinal cord like in the adult rat were found at P7 (Curfs et al. 1996). Myelination of the pyramidal tract starts in the caudal medulla oblongata at P7 (Gorgels 1990, 1991). Myelination of corticospinal axons in the spinal cord begins rostrally (C5) at about P14 and continues caudalwards during the third postnatal week (Joosten et al. 1989). A close temporal relationship exists between the appearance of the fore- and hindlimb placing responses and the arrival of corticospinal axons in the spinal gray matter (Donatelle 1977). Forelimb placing is first seen between P4 and P7, and hindlimb placing between P9 and P13.

Which mechanisms control fiber outgrowth into the corticospinal spinal gray target areas? A diffusible chemotropic signal could be one of the environmental cues involved in axonal outgrowth and guidance. By using the collagen gel co-culture technique (Lumsden and Davies 1983), the basilar pons was shown to become innervated by controlling the budding and directed outgrowth of corticospinal axon collaterals through the release of a diffusible chemotropic substance (Heffner et al. 1990; O'Leary et al. 1990, 1991). Similarly, it is likely that the cervical spinal gray matter becomes innervated by corticospinal axons through the release of a diffusible chemotropic factor (Joosten et al. 1991). The cell adhesion glycoprotein L1 may be

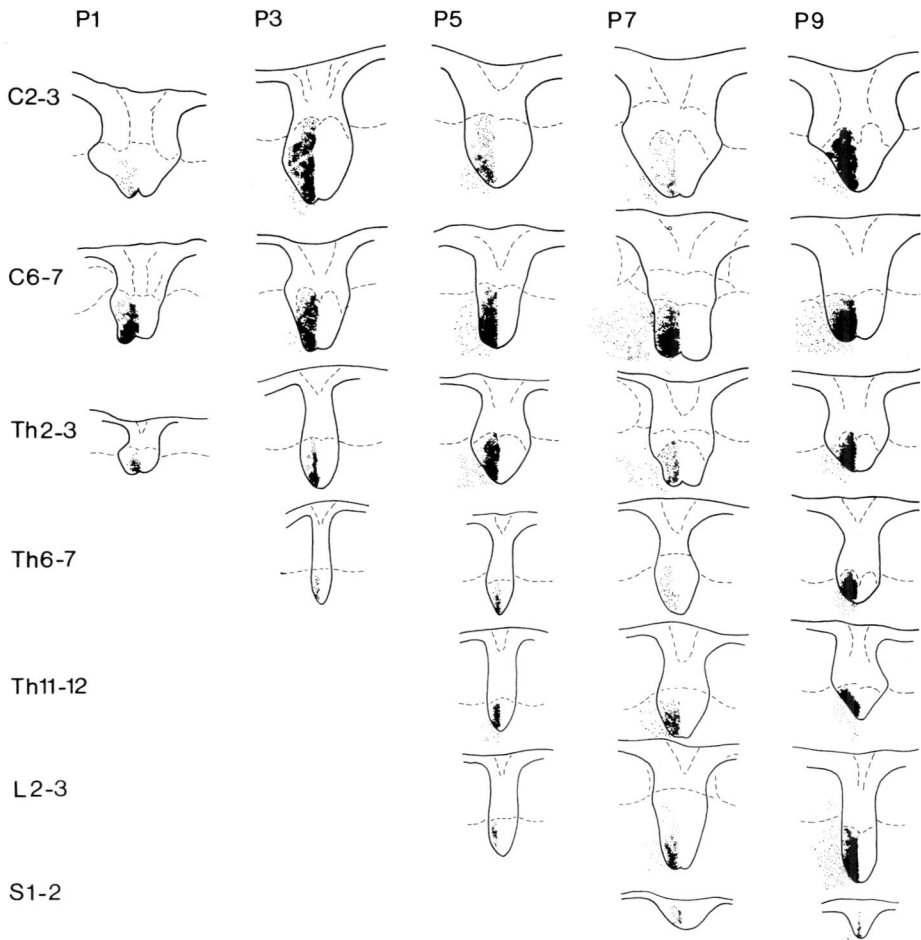

Fig. 39. The outgrowth of the corticospinal tract into the spinal cord of P1–P9 rats. The corticospinal tract was labeled with WGA-HRP. (After Gribnau et al. 1986)

involved in fasciculation of outgrowing later arriving corticospinal fibers (Joosten et al. 1990). It is not involved in the onset of myelination of the corticospinal tract. In L1 mutant mice, however, the L1 mutation causes a primary pathfinding deficit in the development of the corticospinal decussation rather than errors of fasciculation by late growing axons (Cohen et al. 1997, 1998; Dahme et al. 1997). The likely interaction that is disturbed is the pathfinding by corticospinal axons along the previously grown dorsal column projection. During the outgrowth of the corticospinal tract, Joosten and Gribnau (1989) noted a prominent vimentin-immunoreactive glial septum in the midline raphe of the hindbrain and spinal cord. Such a glial septum is absent in the decussation area of corticospinal tract fibers. This glial septum may act as a physical barrier during the outgrowth of the corticospinal tract by preventing its decussation. Oligodendrocyte-associated neurite growth inhibitors in the already myelinated cuneate and gracile fascicles play an important role in channeling and "guard-rail" function for the developing corticospinal tract in the rat spinal cord (Schwab and Schnell 1991).

Descending pathways in the embryonic rodent CNS have the capacity to grow through a lesion and restore conduction (Saunders et al. 1992; Iwashita et al. 1994). This capacity for regrowth remains during only a small period of time early in development. In neonates, restructuring of corticospinal connectivity occurs after unilateral pyramidotomy or lesion of the sensorimotor cortex (Kalil and Reh 1982; Reh and Kalil 1982b; Bernstein and Stelzner 1983; Schreyer and Jones 1983; Rouiller et al. 1991). The immunological disruption of mature myelin or suppression of myelin inhibitory proteins within the rodent spinal cord facilitates CNS axonal regrowth and/or sprouting (Schnell and Schwab 1990, 1993; Bregman et al. 1995; Dyer et al. 1998; von Meyenburg et al. 1998; Z'Graggen et al. 1998). Treatment with the myelin-specific antibody IN-1 facilitates the regeneration of rubrospinal axons after a lateral hemisection (Dyer et al. 1998), and functional recovery of forelimb use occurs after pyramidal tract lesion (Z'Graggen et al. 1998). Fetal spinal cord transplants also contribute to recovery of function by restoring some supraspinal input (e.g., Bregman and Bernstein-Goral 1991; Bernstein-Goral and Bregman 1993; Joosten 1997; Diener and Bregman 1998a,b). Axonal regrowth contributes not only to recovery of rhythmic alternating movements such as stepping, but also to skilled forelimb activity such as reaching and grasping (Diener and Bregman 1998a,b). Following a high cervical spinal cord lesion in adult rats, levels of the inducible transcription factor c-Jun and of the growth-associated protein GAP-43 in rubrospinal tract neurons are greatly elevated (Jenkins et al. 1993; Tetzlaff et al. 1994; Kobayashi et al. 1997; Houle et al. 1998). Regeneration-related genes are turned on best when the axotomy is close to the cell body. A thoracic level injury fails to upregulate c-Jun expression.

8.3
Tract-Tracing Data in Cats

Data on the development of descending supraspinal pathways in cats are limited, and mainly focussed on the development of corticospinal and corticorubral projections. Bruce and Tatton (1981) found that corticospinal, rubrospinal, reticulospinal and vestibulospinal projections are all well developed at P20. The classical retrograde degeneration studies by Brodal and co-workers (Pompeiano and Brodal 1957a,b; Torvik and Brodal 1957), however, showed that the full complement of reticulospinal, vestibulospinal, and rubrospinal projections is already well developed in young kittens (P6 and beyond). Moreover, after hemisection of the thoracic cord in newborn kittens, Bregman and Goldberger (1983c) found retrograde cell changes in the red nucleus and the lateral vestibular nucleus suggesting that the rubrospinal and lateral vestibulospinal tracts have obviously extended into the lumbar spinal cord during gestation. Serotonergic raphespinal projections are present throughout the spinal cord at birth (Goldberger 1986). Noradrenergic projections are present but not so well developed as the dense serotonergic projections. Windle's classical studies suggest that descending brainstem pathways are formed between E16, when the rostral component of the flm is found (Windle 1932b), and roughly E32 when the fetal period starts. Windle and Griffin (1931) studied the development of motility from its early manifestation in the 16-mm embryo (~E24) to a time shortly before birth. The earliest motility noted was movement of the trunk involving the head region. Forelimb motility appeared first (~E25) as part of trunk movements. The elbow was involved at E26, wrist bending was observed at E30, adduction of the forelimb at E32, and definite digital flexion at ~E42. Hindlimb motility occurred in a similar sequence but at a later time (from E28 till E47).

Wise et al. (1977) showed that corticospinal fibers begin to grow into the cervical spinal cord 12 to 15 days prior to birth (E65=P0), i.e., before there was any sign of laminar organization of the cortex and before other corticofugal projections such as corticocollicular or corticopontine tracts could be recognized. They suggested that this early appearance was followed by a "waiting" period before the axons finally invaded their zones of termination. At birth, corticospinal fibers have reached all levels of the spinal cord (Leonard and Goldberger 1987b; Theriault and Tatton 1989). Oka et al. (1985) showed that kitten corticospinal tract neurons do undergo changes postnatally and do not show a mature form of responsiveness until around 4 weeks after birth. Alisky et al. (1992) showed that there are three phases in the spatial development of corticospinal projections (Fig. 40): (1) an early period (P1–P10) when corticospinal axons invade the spinal gray from the white matter; (2) an intermediate period (2–5 postnatal weeks) where these axons develop terminal arborizations in a rostral to caudal, medial to lateral and intermediate gray to dorsal and ventral horn sequence; and, (3) a late period (6–7 postnatal weeks) during which some corticospinal projections are eliminated. Selective elimination of the transient ipsilateral projections to the dorsal horn and the dorsolateral part of the intermediate zone, and of the bilateral projections to the ventral horn was found, leaving an adult-like pattern of termination (see Fig. 40). In cats, myelination of the corticospinal tract is still incomplete at 1 month of age, and proceeds slowly to completion at 4–5 months after birth (Huttenlocher 1970).

In DiI experiments, McConnell et al. (1994) showed that corticofugal axons invade the internal capsule by E26, and reach the pons before E55. Song et al. (1995a) found that corticorubral axons invaded the red nucleus by E50, whereas cerebellorubral fibers entered the red nucleus already before E35. A similar developmental sequence of rubral afferent projections was found in the North American opossum (Martin et al. 1986, 1988). Probably, cat rubrospinal axons innervate the spinal cord before E35. Myelination of cat rubrospinal axons starts prior to E59 (Song et al. 1995b). Song and Murakami (1998) studied the development of the pattern of functional cortical inputs to individual rubrospinal neurons: in preterm kittens (E61-E65), only about half of the rubrospinal neurons showed adult-like functional topography, to increase to 82% in P1–P8 kittens and 93% in P13–P28 kittens. These data as well as tract-tracing data (Higashi et al. 1990) suggest that corticorubral axons make functional synapses nonselectively with rubrospinal neurons before birth.

Bregman and Goldberger (1983a–c) showed that kittens subjected to spinal cord damage have a considerable capacity for reorganization and rerouting of corticospinal fibers, and that this is probably reflected in the recovery of some motor functions such as tactile placing. In contrast, descending brainstem pathways which develop earlier than the corticospinal tract showed little or no recovery, and the motor patterns which these subserved (e.g., postural reflexes and locomotion) also failed to

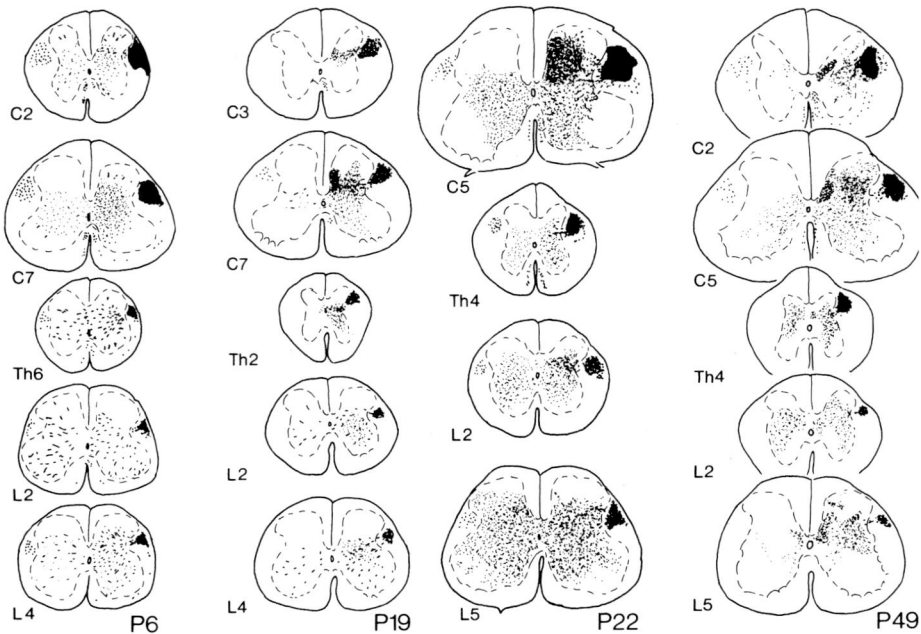

Fig. 40. The development of the corticospinal tract in cats at P6, P19, P22, and P49, when an adult-like pattern of labeling is present. (After Alisky et al. 1992)

recover. Leonard and Goldberger (1987a,b) studied the consequences of damage to the sensorimotor cortex in neonatal and adult cats. The emergence of motor behavior during development in neonatal operates appeared to follow the same pattern as in normal animals although with a protracted time course. Moreover, motor behavior did not achieve the level of maturity of normal animals. Neonatal operates exhibited greater recovery and sparing than adult operates. The anatomical basis for sparing of contact placing is formed by the maintenance of exuberant projections present in normal neonatal animals. As in normal newborn cats, the intact sensorimotor cortex of neonatal operates exhibited bilateral corticothalamic and corticorubral projections not present in normal or adult operated cats. Dense bilateral corticothalamic and corticorubral projections were also found in normal newborn animals. The crossed corticothalamic and corticorubral projections are supposed to play a role in sparing and recovery of function, particularly in sparing of contact placing (Leonard and Goldberger 1987b).

8.4
Tract-Tracing Data in Primates

Data on the development of descending supraspinal pathways in primates are restricted to immunohistochemical and time of neuron origin data on monoaminergic pathways and neuroanatomical and functional data on the corticospinal tract. No experimental data are available on the time of origin and outgrowth of descending brainstem pathways. Bodian (1966, 1968) studied a series of macaque (*Macaca irus* and *M. mulatta*) embryos (estimated ovulation age 42–51 days; 17–41 mm CRL) comparable to stages 21–23 of Gribnau and Geijsberts (1981) subjected to behavioral analysis with the umbilical cord intact, and light and electronmicroscopic analysis of the cervical spinal cord afterwards. Three behavioral stages were distinguished: (1) stage 1 (17–22 mm CRL, stage 21), a pre-reflex group; (2) stage 2 (24–28 mm CRL, stage 22) with early local spontaneous and reflex movements (trigeminal, cervical, brachial); and (3) stage 3 (32–41 mm CRL, stage 23 and beyond), a group characterized by the development of more vigorous activity. These data suggest that supraspinal input must be present at the end of the embryonic period. At stage 23, a primordial red nucleus could be recognized (Gribnau and Geijsberts 1985), and the superior cerebellar peduncle was found decussating in the mesencephalon at stage 22 (Kappel 1981). These observations make it likely that the rubrospinal tract will reach the spinal cord at the end of the embryonic period (E46–50).

In the rhesus monkey, Levitt and Rakic (1982) studied the time of neuron origin and differentiation of brainstem monoamine neurons. Neurogenesis in the locus coeruleus is between E27 and E36 with peak production around E30-E33. In the medial part of the locus coeruleus the majority of neurons is generated on E30, whereas most cells of its lateral part are generated on E32 and E33. Neurogenesis of the serotonergic raphe nuclei occurs between E28 and E43 with only a moderate rostrocaudal spatiotemporal gradient: neurons of raphe nuclei with ascending projections (raphe dorsalis and central superior nuclei) undergo final mitosis between E28 and E35, with a peak on E35; neurons of the raphe magnus, pontis, obscurus and pallidus nuclei, i.e., raphe nuclei with mainly descending projections, are produced between E35 and E43, with peak production between E38 and E40. These data suggest that monoaminergic projections will innervate the spinal cord well before the end of the embryonic period.

In the rhesus monkey, corticospinal fibers have reached at least to the level of the lower cervical segments at birth (Kuypers 1962). Recently, Killackey et al. (1997) studied the distribution of corticospinal projection neurons in the neocortex of a fetal macaque monkey receiving a Fast Blue injection below the pyramidal decussation at E95. At E108, the fetus was sacrificed and labeled neurons were appropriately located in layer Vb, but in a wider area than observed in neonatal monkeys. Similar observations were made by Galea and Darian-Smith (1995) in a monkey delivered about two weeks prematurely (~E158) by cesarian section followed by a cervical Fast Blue injection two days later. Two important differences were noted in the distribution of corticospinal neurons as compared to mature macaques: (1) the areal extent of the corticospinal neurons in the infant monkeys was greater than in the mature macaques, and (2) the local soma density was greater in the infants. Both the areal extent of the cortical origin and the relative number of corticospinal neurons with spinal axons regress very substantially over a period of 2 years (Galea and Darian-Smith

1995). The direct corticomotoneuronal projections do not appear to develop until 6–8 months of age (Fig. 41). Lawrence and Hopkins (1976) extensively studied the development of hand and finger movements in infant rhesus monkeys. The earliest signs of reaching were found at 3–4 weeks of age. Reaching was inaccurate and grasping of food was part of a rather gross whole arm and hand movement. Smooth reaching occurred in the third month and the first signs of relatively independent finger movements were present in the second and third month. Fully mature relatively independent finger movements were present at 7–8 months of age. This developmental time course correlates well with the appearance of corticomotoneuronal projections (Kuypers 1962; Armand et al. 1994, 1996, 1997; Galea and Darian-Smith 1995). In monkeys pyramidotomized at birth there was no appreciable difference in the development of general motor activity. They could run, walk, climb, and jump as well as normal infant animals. Reaching developed in the normal fashion and became smooth and accurate. However, none of the pyramidotomized animals ever developed any relatively independent finger movements (Lawrence and Hopkins 1976; Galea and Darian-Smith 1997b). The maturation of the monkey corticospinal tract was also studied using transcranial magnetic stimulation of the motor cortex (Flament et al. 1992a,b; Olivier et al. 1997). A correlation with the development of relatively independent finger movements was found, and a staggered development of corticospinal projections to forelimb and hindlimb was suggested. Cortically-evoked responses in hand muscles could be recorded about one month earlier than those in foot muscles.

Following extensive unilateral lesions of the cervical spinal cord in the macaque monkey, a remarkable recovery of hand function was found (Galea and Darian-Smith 1997a,b). In newborn and juvenile macaques, spinal sections were made and the operated animals were followed for up to 150 weeks. The origin, trajectory and site of termination of corticospinal pathways were examined at intervals during the period of recovery of hand function. Immediately after unilateral section of the cervical spinal cord (C3), a profound reduction in the corticospinal projection to the hemicord caudal to the lesion was found. The few labeled corticospinal axons spared by the

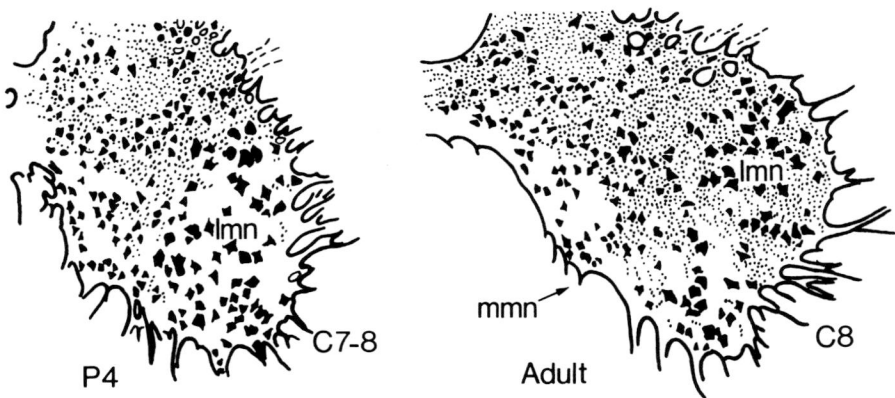

Fig. 41. The distribution of corticospinal terminations in the ventral horn of the rhesus monkey in **A** at 4 days of age, and in **B** in an adult. *lmn, mmn,* lateral and medial motoneurons, respectively. (After Kuypers 1962)

lesion by-passed the lesion by descending in the contralateral cord and then crossing the midline caudal to the lesion. In every monkey, a similar, profound reduction in the corticospinal and rubrospinal projections to the ipsilateral side of the cord caudal to the lesion was observed. This pattern did not alter significantly over an extended recovery period, and it was concluded that there was no substantial reconstruction of the corticospinal projection over a recovery period of more than 2 years (Galea and Darian-Smith 1997a). The remarkable, but incomplete recovery of dexterity over a period of 6–12 months that was observed (Galea and Darian-Smith 1997b) must be achieved by optimizing the transmission of information from the cerebral cortex to the spinal cord by the substantially reduced populations of corticospinal neurons and corticobulbospinal projections and/or the effective use of spinal circuitry in regulating the more stereotyped elements of the manual task.

9 Development of Descending Supraspinal Pathways in Man

9.1
Some Notes on the Development of the Human Spinal Cord

In the spinal cord of pyridine silver stained human embryos of 5–8 weeks of estimated menstrual age, i.e., about 3–6 postovulatory weeks, Windle and Fitzgerald (1937) noted that motoneurons are the first neurons to develop (Fig. 42). They appear in the uppermost spinal segments at approximately E27 (about Carnegie stage 13/14). At this time of development also dorsal root ganglion cells are present. Central processes of the bipolar ganglion cells reach the spinal cord, where they initiate the formation of the dorsal funiculi. At first, the dorsal funiculi are found only in the cervical spinal cord, and are composed of short fibers, but at stage 15 dorsal funiculi are found throughout most of the spinal cord. At stage 18, collateral branches of primary afferent fibers emerge from the lateral aspect of each dorsal funiculus in the brachial region. A few long collateral branches pass into the lateral division of the ventral horn at stage 20. At this stage of development the ventral funiculus may contain descending axons from the brain stem passing via the flm. Most of its other fibers are probably ascending, however (Rhines and Windle 1941). Interneurons with ascending projections send their axons to the floor plate where they cross in the ventral commissure and form contralateral ascending tracts in the ventral funiculus. Therefore, three components of cutaneous reflex pathways (primary afferent fibers, interneurons and motoneurons) are already found in a human embryo of 4 postovulatory weeks. A rapid differentiation of these components takes place in embryos of 6 postovulatory weeks (see Fig. 42). The dorsal funiculus has reached the caudal brain stem at stage 16, i.e., about 37 postovulatory days (Müller and O'Rahilly 1989a). Cuneate and gracile decussating fibers forming the medial lemniscus are present at stage 20 (Müller and O'Rahilly 1990a,b). Three types of primitive sensory neurons, presumably transient structures, were described by Humphrey (1944, 1947). Human embryos are capable of movement by the time they have attained a length of 20–21 mm, i.e., at about 7.5 weeks estimated menstrual age or about 5.5 postovulatory weeks (Fitzgerald and Windle 1942; Hooker 1952; Humphrey 1964).

Konstantinidou et al. (1995) studied the development of the primary afferent projections in the fetal human spinal cord between 8 and 19 weeks of gestation (about 6 to 17 postovulatory weeks) using DiI tracing. They confirmed Windle's work by showing that dorsal root fibers enter the spinal gray matter very early in development. By 6 postovulatory weeks, a few axons (presumably muscle spindle afferents) already

95

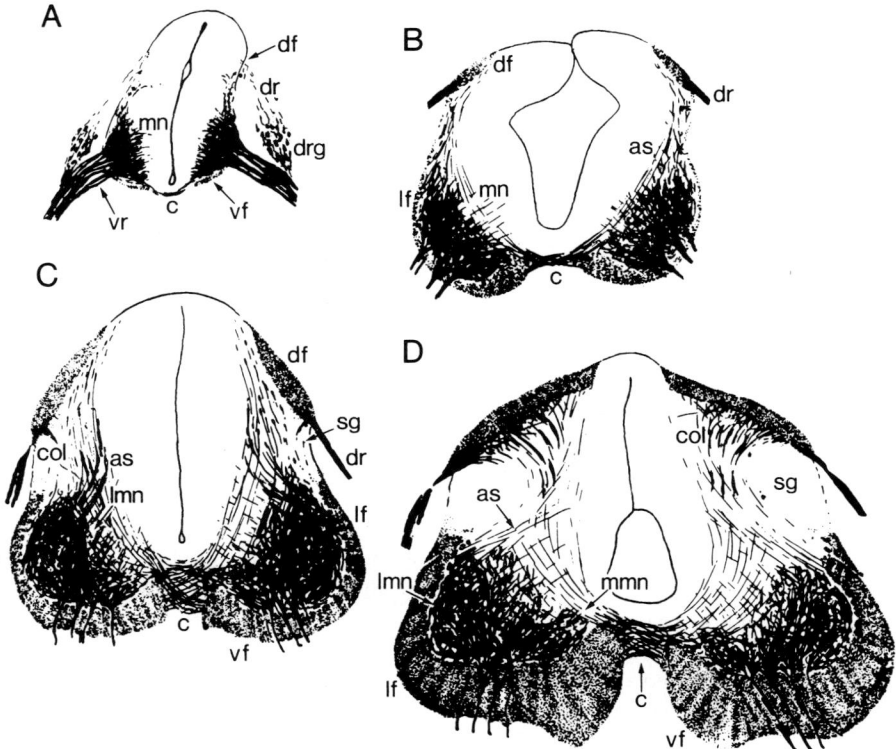

Fig. 42. Windle and Fitzgerald's data obtained with silver staining in human embryos of 5–8 weeks of gestation. *as*, association interneurons; *c*, ventral commissure; *col*, dorsal root collaterals; *df*, dorsal funiculus, *dr*, dorsal root; *drg*, dorsal root ganglion; *lf*, lateral funiculus; *lmn*, *mmn*, lateral and medial motoneurons, respectively; *mn*, motoneurons; *sg*, substantia gelatinosa; *vf*, ventral funiculus; *vr*, ventral root. (After Windle and Fitzgerald 1937)

reached the motor pools. As development progresses, these axons project to the ventral horn and branch in a restricted area in the intermediate zone as well as in the motor pools. Between 9 and 17 postovulatory weeks, axon collaterals in the ventral horn form boutons in the proximity of motoneuron somata and their proximal dendrites. Other groups of axons penetrate the spinal gray matter via the mediolateral extent of the dorsal horn to reach lamina IV, and then turn upward to terminate in layers III and IV. Probably, these axons arise from dorsal root ganglion cells that innervate low-threshold mechanoreceptors.

The synaptogenesis in the lateral motor column of the cervical spinal cord was studied by Okado and co-workers (Okado et al. 1979; Okado 1980, 1981). The first synapses were found in the motor nucleus of the cervical cord in a 10-mm embryo, i.e., at Carnegie stage 15 (estimated ovulation age: 30–32 days). Since no primary afferent fibers extend far enough to reach the motor neuropil, these axodendritic synapses probably come from interneurons. The first synapses between primary afferent fibers and interneurons were found in a 14-mm embryo belonging to

Carnegie stage 17. The first axosomatic synapses were found in the motor neuropil at stage 17. Okado's data suggest that the formation of synapses between interneurons and dendrites of motoneurons precedes that of synapses between interneurons and collaterals of dorsal root fibers. During the first 5 months of development there appear to be three critical periods of synaptogenesis coinciding with behavioral changes found in human fetuses (Okado and Kojima 1984): (1) a period of closure of the spinal reflex arc, i.e., the onset of synapse formation, coincides with the appearance of spinal reflex activities; (2) a period of rapid increase of axodendritic synapses that corresponds with the onset of local activities (Humphrey 1964); and (3) a period with an increase of axosomatic synapses. Using real-time ultrasound, de Vries et al. (1982, 1984) found the first discernible spontaneous movements of the fetus at 7.5 weeks of gestation, i.e., about 5.5 postovulatory weeks or approximately stage 16. By the end of the embryonic period, the following types of prenatal movements are discernible by ultrasound: startles, general movements, hiccups, isolated limb movements, head retroflexion and rotation, and hand-face contact. Such movements reflect coordinated motor patterns (de Vries et al. 1982, 1984). By this time, descending supraspinal pathways arising in the interstitial nucleus of the flm, the reticular formation of the brain stem and the vestibular nuclear complex must have reached the spinal cord.

9.2
Development of Descending Brainstem Pathways in Man

Assuming that the stages of neural development are similar in rats and man even though their exact chronological ages are different, Bayer et al. (1995) estimated human neurogenetic time-tables by extrapolating the rat data (see Sect. 8.2) to the longer span of human development. Most brainstem nuclei innervating the spinal cord are born between 4 and 7 weeks after fertilization (see Table 8). The first descending brainstem projections to the spinal cord in the human embryonic brain arise in the interstitial nucleus of the flm and in the reticular formation. At early developmental stages (from stage 11/12 onwards), in the brain stem a ventral longitudinal tract can be distinguished, followed by lateral and medial longitudinal fasciculi at stage 13. Neurofibrillar differentiation is first found in the interstitial nucleus (Rhines and Windle 1941; Windle 1970). Descending fibers from the medullary reticular formation reach the spinal cord in embryos of 10–12 mm CRL (Windle and Fitzgerald 1937). Interstitiospinal fibers from the interstitial nucleus of the flm start to descend in the flm at stage 13, i.e., at E28 (Müller and O'Rahilly 1988a,b). At stage 14, the flm joins the ventral longitudinal fascicle which originally is composed of ascending fibers (Fig. 43). Tectobulbar tracts join the lateral longitudinal fascicle at stage 16 (Windle 1970; O'Rahillly et al. 1987). In 12-mm CRL embryos (about stage 17/18), vestibulospinal projections were found (Windle 1970). The red nucleus can first be recognized in stage 17 embryos (Cooper 1946; O'Rahilly et al. 1987; Müller and O'Rahilly 1989b), and the brachium conjunctivum differentiates at stage 19 (Cooper 1946; O'Rahilly et al. 1988). At the end of the embryonic period, the flm is well-developed, and receives ascending and descending (the medial vestibulospinal tract) components from the vestibular nuclear complex (Müller and O'Rahilly 1990c). The lateral vestibulospinal tract arises from the lateral vestibular nucleus, and is composed of coarse fibers.

Table 8. The estimated time of development of descending supraspinal pathways in man

Nuclei	Estimated time of neuron origin (postovulatory weeks)[a]	Estimated innervation of the spinal cord (Carnegie stages)
Reticular formation		
Medullary	4.1–7.0	~14/15[g]
Pontine	4.1–7.0	
Mesencephalic	5.3–7.0	
Interstitial nucleus of flm	4.1–5.7 (related Darkschewitsch nucleus)	14?[c,f]
Raphe nuclei	3.5–7.0	
Serotonergic projections		?
Vestibular nuclei		
Lateral vestibular nucleus	4.1–5.7	~17/18[f]
Medial and inferior vestibular nuclei	4.1–7.0	Before end of embryonic period[d]
Locus coeruleus		
Coeruleospinal neurons	Peak 4.1–5.2	
Noradrenergic projections		18[e]
Red nucleus	5.3–6.6	?
Hypothalamus		
Paraventricular nucleus	5.3–7.0	?
Corticospinal projections (layer V neurons)	(5.8) 6.7–9.9	Early fetal period[b]

[a] Bayer et al. 1995.
[b] Humphrey 1960.
[c] Müller and O'Rahilly 1988a, b.
[d] Müller and O'Rahilly 1990c.
[e] Puelles and Verney 1998.
[f] Windle 1970.
[g] Windle and Fitzgerald 1937.

Monoaminergic projections also appear to arise early in human development. In 8 mm CRL embryos (about stage 14), Windle and Fitzgerald (1942) observed that descending fibers from the presumptive locus coeruleus join the lateral longitudinal fascicle. A definite locus coeruleus can be distinguished at stage 17 (Müller and O'Rahilly 1989b). Olson et al. (1973) detected neurons containing catecholamines or serotonin in embryonic brains as early as 8 weeks (21 mm CRL), and some catecholaminergic medullary neurons with spinal projections in 10-week fetuses (40 to 45 mm CRL). Nobin and Björklund (1973) first identified the locus coeruleus in 3-month-old fetuses. Using antibodies against enzymes of the catecholamine pathway, noradrenergic groups were labeled in the medulla oblongata and in the locus coeruleus as early as 6 weeks of gestational age (Verney et al. 1991; Zecevic and Verney 1995). Recently, Puelles and Verney (1998) studied the early neuromeric distribution

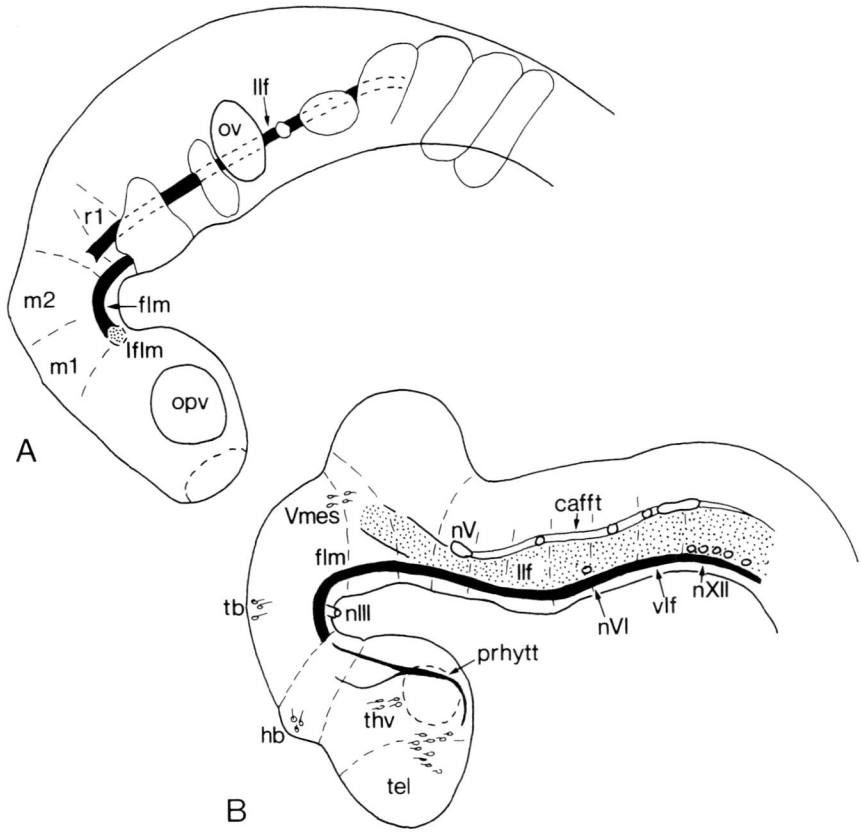

Fig. 43. The extent of the flm in stage 13 (**A**) and 14 (**B**) human embryos. *cafft*, central afferent tract; *flm*, fasciculus longitudinalis medialis; *hb*, habenula; *Iflm*, interstitial nucleus of flm; *llf*, lateral longitudinal tract; *m1*, *m2*, mesomeres 1 and 2, respectively; *nIII*, *nV*, *nVI*, *nXII*, cranial nerves; *opv*, optic vesicle; *ov*, otic vesicle; *prhytt*, pretectohypothalamic tract; *r1*, rhombomere 1; *tb*, tectobulbar tract; *tel*, telencephalon; *thv*, ventral thalamus; *vlf*, ventrolateral fascicle; *Vmes*, mesencephalic nucleus of the trigeminal nerve. (After Müller and O'Rahilly 1988a,b)

of tyrosine hydroxylase (TH)-immunoreactive neurons in human embryos in stages 15–18, i.e., between 4.5 and 6 postovulatory weeks (for a discussion of neuromeres in staged human embryos see Müller and O'Rahilly 1997). In the youngest embryo (4.5 postovulatory weeks, 10 mm CRL, stage 15), TH-immunoreactive neurons were already present in the isthmic neuromere. Descending catecholaminergic fibers were found in the spinal cord at stage 18.

9.3
Development of Corticospinal Projections in Man

The corticospinal tract is one of the latest developing descending pathways. At stage 21, the cortical plate starts development, whereas the definite internal capsule is present by stage 22 (Müller and O'Rahilly 1990b). Hewitt (1961) found the earliest sign of the internal capsule (probably the thalamocortical component) in stage 18 (13 –17 mm CRL). The development of the human corticospinal tract was examined by Humphrey (1960) with a silver technique (Fig. 44). The pyramidal tract reaches the level of the pyramidal decussation at the end of the embryonic period, i.e., at 8 weeks of development (Müller and O'Rahilly 1990c). Pyramidal decussation is complete by 17 weeks gestational age, and the rest of the spinal cord is invaded by 19 (lower thoracic cord) and 29 (lumbosacral cord) weeks gestational age (Humphrey 1960). Myelination of the pyramidal tract is already in progress at the level of the pyramidal

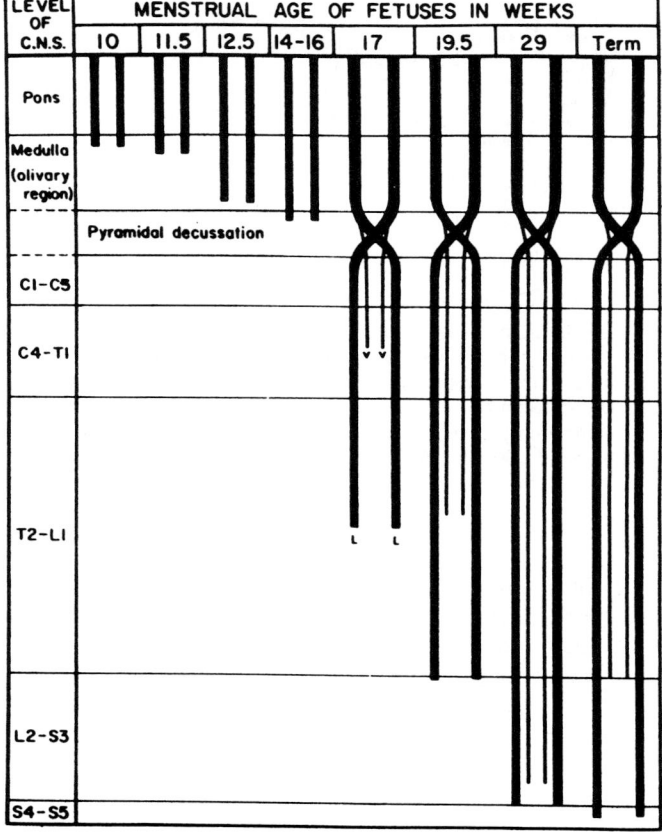

Fig. 44. The outgrowth of the human corticospinal tracts. *L*, lateral corticospinal tract; *V*, ventral corticospinal tract. (After Humphrey 1960)

decussation in a 220 mm CRL fetus at 25 weeks of gestation (Woźniak and O'Rahilly 1982). Myelination of the pyramidal tract occurs over a protracted period (Yakovlev and Lecours 1967).

Early myelination of the corticospinal tracts and other spinal pathways was extensively studied using antibodies against myelin-associated proteins, myelin basic protein in particular (Tohyama et al. 1991; Weidenheim et al. 1992, 1993, 1996; Bodhireddy et al. 1994). Fiber tracts that appear early in development generally undergo myelination before later appearing tracts (Yakovlev and Lecours 1967; Gilles et al. 1983; Brody et al. 1987; Kinney et al. 1988). In the brain stem, myelination starts in the flm at 8 postovulatory weeks. The vestibulospinal tracts become myelinated at the end of the second trimester, whereas the pyramidal tracts begin very late (at the end of the third trimester) and myelination is not completed in them until about 2 years. The temporal and spatial expression of myelin basic protein (MBP) in the first and early second trimester of the human spinal cord is shown in Fig. 45. MBP is expressed in a rostral-to-caudal and anterolateral-to-posterior manner in most tracts of the spinal cord. In the fasciculus gracilis, however, myelination starts at the lumbar level. In the fetal spinal cord, in MBP-stained sections the corticospinal tracts stand out as unstained areas in the white matter.

In man, the maturation of skilled finger movements requires a much longer period of development than in the rhesus monkey (Forssberg et al. 1991). This maturation is also dependent on that of the corticospinal tracts (Eyre et al. 1991; Müller et al. 1991,

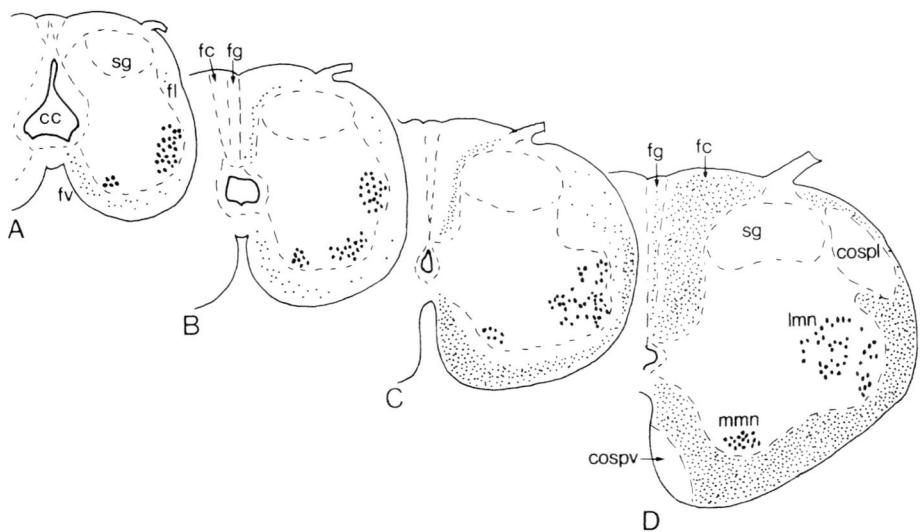

Fig. 45A–D. The development of myelination in the early human spinal cord as found by the expression of myelin basic protein. **A–D** Transverse sections through the cervical spinal cord of specimens 10, 12, 16 and 23 gestational weeks of age, respectively. *cc*, central canal; *cospl, cospv*, lateral and ventral corticospinal tracts, respectively; *fc*, fasciculus cuneatus; *fg*, fasciculus gracilis; *fl, fv*, lateral and ventral funiculi, respectively; *lmn, mmn*, lateral and medial motoneuron columns, respectively; *sg*, substantia gelatinosa. (Based on data by Weidenheim et al. 1993, 1996, and Bodhireddy et al. 1994)

1997). For the first two years of life, a rapid decline was shown in the central conduction time of responses to magnetic stimulation of the cortex in humans aged from 32 weeks gestation to 55 years. Adult values for central conduction times in both biceps brachii and hypothenar muscles were achieved around 2–4 years of age. This extended time course is in keeping with the protracted period during which myelination of the human pyramidal tract continues.

Although developmental disorders will be found in most if not in all descending supraspinal pathways, in man aplasia and anomalies of the corticospinal tracts have received most attention. Developmental disorders in the pyramidal tracts can occur in various stages of development. Bilateral absence of the pyramids in the brain stem has been found in the dorsal (*anencephaly*) and ventral (*holoprosencephaly*) induction stages of human development, in antenatal and perinatal destructive lesions such as schizencephaly and porencephaly, in X-linked hydrocephalus, in microcephaly and in neuronal migration disorders (Chow et al. 1985; Friede 1989; Aicardi 1992; Norman et al. 1995; ten Donkelaar et al. 1999). Several variations in the funicular trajectory of the pyramidal tracts were described in more or less normally developed cases (Verhaart and Kramer 1952; Nyberg-Hansen and Rinvik 1963; Yakovlev and Rakic 1966; Nathan et al. 1990). Anomalies in the crossing of the corticospinal tracts are often associated with encephaloceles (Verhaart and Kramer 1952), and vermis aplasia in Dandy-Walker's and Joubert's syndromes (Friede and Boltshauser 1978; Lagger 1979; ten Donkelaar et al., in preparation). Developmental disorders of the cerebellum may include absence of the vestibular nuclei, and consequently of the vestibulospinal tracts (Friede 1989). Like in mice, the neural cell recognition molecule L1 appears to be required for the formation of the corticospinal tracts in humans (Brümmendorf et al. 1998; Kamiguchi et al. 1998). Mutations in the L1 gene underlie a highly variable recessive neurological disease characterized by mental retardation, spasticity and flexion deformities of the thumbs and varying degrees of brain malformation such as hydrocephalus, hypoplasia of the corticospinal tracts and an underdeveloped corpus callosum (Kenwrick et al. 1996; Fransen et al. 1997). Congenital white matter hypoplasia, a progressive familial disease that presents in childhood with severe hypoplasia of the cerebral white matter and absence of the pyramids in the medulla (Chattha and Richardson 1977; Lyon et al. 1990), may be due to a pathological extension of the normal axonal elimination (Lyon et al. 1990).

10 Concluding Remarks

In this final section some major trends in the ontogenesis of descending supraspinal pathways in tetrapods will be summarized.

10.1
The Outgrowth of Descending Supraspinal Pathways Is the Result of a Coordinated Program

The development of the CNS proceeds in a series of decisions made by neuroectodermal precursors and their progeny, decisions that outline the fate of a differentiated neuron including its pattern of axonal connections. Like for other pathways in the developing CNS, the outgrowth of axons of descending supraspinal pathways can be regarded as the result of a series of distinct processes, which may be expressed in a coordinated program (Jacobson 1991): (1) the outgrowth of axons and selection of pathways to their appropriate destination; (2) dendritic outgrowth and formation of specific dendritic morphology; (3) selection of specific targets and collateralization by axons; (4) elimination of incorrect and redundant synapses, axonal and dendritic branches, and of mismatched neurons; and (5) functional refinement of synaptic connections.

Shortly after neurons are "born," i.e., have undergone their final mitosis, a growth cone arises followed by axonal elongation and the formation of dendrites. By injecting HRP into *Xenopus laevis* blastomeres, Jacobson and Huang (1985) were able to follow the differentiation of the descendant cells. The tracer entered the outgrowing neurites and showed the position and directions of initial outgrowth of axons as well as of dendrites, the pattern of neuritic branching, and the relation between axon terminals and specific targets. All types of nerve fibers studied grew by the most direct pathway, apparently without errors of initial outgrowth, pathway selection, or target selections. Transmitter immunohistochemistry (van Mier et al. 1986; Gallagher and Moody 1987; Roberts et al. 1987, 1988) in *X. laevis* also clearly showed that from the moment cell division stops, an axon is formed followed by dendrites which emerge from the cell body. At the beginning of the differentiation phase the production of the cell-specific neuroactive substances takes place. Initial outgrowth is in a specific direction for each class of neuron (Roberts 1988). It is likely that all descending supraspinal pathways arise in a similar way.

In the spinal projections of each of the descending supraspinal pathways three stages can be distinguished: (1) an initial stage of outgrowth to the spinal cord; (2) a short "waiting" period after which collaterals enter the spinal gray matter; and (3) myelination of axons. This has been demonstrated for raphespinal fibers (van Mier et al. 1986; Okado et al. 1992), reticulospinal fibers (Shiga et al. 1991) and corticospinal projections (e.g., O'Leary and Koester 1993). In chicken embryos, Shiga et al. (1991) observed a delay of about two days between the arrival of supraspinal fibers in the white matter and the penetration of the gray matter. Since descending supraspinal fibers establish axodendritic synapses in the white matter half a day earlier, the actual delay in this case is about one and a half days. That a delay between the arrival of supraspinal fibers in the spinal white matter and the penetration of the gray matter is also obvious for other descending pathways is illustrated by experiments on the development of the rubrospinal tract (Fig. 34) and noradrenergic, coeruleospinal projections (Fig. 33) in the North American opossum.

In all mammalian species studied so far, the developing corticospinal tract arises from a much larger cortical area than the mature tract. Corticospinal axons first extend out of the cortex toward the spinal cord, bypassing their subcortical targets (O'Leary and Koester 1993). Subsequently, the subcortical targets are exclusively contacted by axon collaterals that develop by branching off the flank of a spinally directed primary axon. During development the density of cortical innervation exceeds that of adult mammals, and moreover a wider distribution of corticospinal fibers to the spinal gray matter is found. Such an "overshoot" of spinal projections is not so evident for other descending supraspinal pathways, but in the North American opossum transient spinal projections were found from certain hypothalamic nuclei. Specific branches and segments of the primary corticospinal axons are selectively eliminated to yield the mature projection pattern. The withdrawal of the transitory spinal collaterals probably accounts for the dramatic loss of fibers from the corticospinal tract during development. In the rhesus monkey, the areal extent of the cortical origin and the relative number of corticospinal neurons with spinal axons regress very substantially over a period of two years (Galea and Darian-Smith 1995). In man, a pathological extension of the normal axonal elimination has been suggested in congenital white matter hypoplasia, a progressive familial disease with severe hypoplasia of the cerebral white matter and absence of the pyramidal tract (Lyon et al. 1990).

10.2
The Pattern of Early Descending Axonal Tracts Is Similar in All Vertebrate Groups

Research in invertebrates (see Goodman 1996; Nassif et al. 1998) has shown that early generated, pioneer neurons play an important role in the establishment of axon tracts. These neurons lay down an axonal scaffold containing guidance cues that are available to later generated growth cones. This axonal scaffold has been labeled with immunohistochemical techniques such as staining against axonal glycoproteins of the Fasciclin family in the developing grasshopper (Bastiani et al. 1987; Boyan et al. 1995) and in embryonic *Drosophila* brain (Goodman and Doe 1993; Nassif et al. 1998). The adhesion molecule Fasciclin II (Fas II) is expressed in a large number of early differentiating neurons and can be used to follow the development of axon tracts of the brain. The Fas II antigen is present on the surface of clusters of neuronal somata prior to axon outgrowth. These "fiber tract founder" clusters (Nassif et al. 1998) are laid out in a linear pattern. After expressing Fas II on their soma, neurons of the fiber tract founder clusters extend axons that grow along the surface of the founder clusters and form a simple system of pioneer tracts for each of the components of the brain neuropile. Moreover, because Fas II expression goes on into the larval period, when the first features of the reorganizing adult brain become evident, it is possible to trace the development of several embryonic pioneer tracts fairly clearly into the corresponding adult pathways. Pioneer axons in the *Drosophila* embryo may be guided by multiple cues including glial cells (Jacobs and Goodman 1989; Hartenstein et al. 1998). The removal of one of these cues, for instance the neuropile glial cells, does not necessarily lead to the total disability of axons to reach their target, but enhances the frequency of error in pathfinding.

There is now enough evidence that similar mechanisms may operate in vertebrates (Easter et al. 1994; Goodman and Tessier-Lavigne 1997). This axonal scaffold can now be labeled with several markers, but was in fact already indicated by Herrick (see Fig. 8), Windle and their respective co-workers using reduced silver staining techniques. Herrick (1938, 1939) clearly described sensory dorsolateral and motor ventrolateral bundles and three transversely oriented tracts. Katz and Lasek (1979, 1981) introduced the terms alar and basal substrate pathways for the two longitudinal tracts. Simple acetylcholinesterase staining in zebrafish (Hanneman and Westerfield 1989; Ross et al. 1992), *Xenopus laevis* (Moody and Stein 1988), and chicken (Weikert et al. 1989), and in particular the application of anti-acetylated tubulin and HNK-1 antibodies in various vertebrates including lampreys (Kuratani et al. 1998), zebrafish (Chitnis and Kuwada 1990; Metcalfe et al. 1990; Wilson et al. 1990; Ross et al. 1992), *X. laevis* (Nordlander 1989, 1993; Hartenstein 1993; Easter et al. 1994), chickens (Chédotal et al. 1995) and mice (Easter et al. 1993) beautifully illustrate how the first axonal tracts are laid down in the embryonic brain and spinal cord. All of these early axons course in very restricted superficial regions, leaving most of the surface of the brain free. This early network of tracts may be conserved across vertebrates, and thus represents the earliest solution to the problem of how to start an axonal network (Easter et al. 1994).

Throughout vertebrates including man, the flm is the first descending pathway to be formed. Interstitiospinal fibers "pioneer" this tract, and are joined by reticulospinal fibers. Vestibulospinal fibers (the medial vestibulospinal tract) follow much later. The lateral vestibulospinal tract takes a separate course through the brain stem. Late-arriving fiber tracts such as the rubrospinal and corticospinal tracts probably have their own mechanism of selecting the appropriate pathway. This is also reflected in the various ways by which the corticospinal tract reaches the spinal cord (see Kuypers 1981). The mechanisms through which descending supraspinal pathways select their pathway are largely unknown. Factors involved include (Glover 1993): (1) the polarity of axon outgrowth and the orientation of this outgrowth relative to nearby pathways; (2) the ability to react to nearby pathways; (3) the decision to choose a particular pathway; and (4) the decision to grow in a particular direction along the chosen pathway.

Considerable progress has been made in understanding the molecular basis of the processes by which axons navigate to their correct targets during the development of the CNS (Tessier-Lavigne and Goodman 1996; Drescher et al. 1997; Goodman and Tessier-Lavigne 1997). Much of this progress is attributable to the identification of families of proteins that are responsible for either promoting or inhibiting growth cone extension. Such proteins may be expressed on the surface of, or secreted by, cells in the vicinity of the growth cone, and include the semaphorins, members of the Eph receptor tyrosine kinase signaling system (ephrins), cadherins and immunoglobulin superfamily members. Other proteins, such as the netrins, may influence growth cone behavior at some distance from their site of production or secretion. Mutant analysis in the zebrafish is rapidly revealing the genes that are expressed in the early neuroepithelium and that regulate factors responsible for the guidance of commissural and other axons (e.g., Karlstrom et al. 1997; Wilson et al. 1997). In mice, mutations in the L1 gene cause a primary pathfinding deficit in the development of the corticospinal decussation. In man, such mutations result in a highly variable recessive neurological disorder characterized by mental retardation, spasticity, and hypoplasia of the corticospinal tracts and corpus callosum (Brümmendorf et al. 1998; Kamaguchi et al. 1998).

10.3
The Formation of Descending Supraspinal Pathways Occurs According to a Developmental Sequence

The phylogenetic constancy of descending supraspinal pathways in tetrapods suggests a comparable pattern of development, i.e., a *developmental sequence*. Table 9 summarizes the time of arrival of these pathways in the rostral spinal cord of terrestrial vertebrates. In all species studied, reticulospinal and interstitiospinal fibers reach the spinal cord first, followed by vestibulospinal fibers and, much later, by rubrospinal and, if present, corticospinal projections. A special case is presented by anurans which in fact have two motor systems, a primary, transient motor system and a secondary, definitive motor system. Reticulospinal, interstitiospinal and vestibulospinal fibers as well as the Mauthner cells innervate the spinal cord very early in development, well before the development of the hindlimbs. Rubrospinal fibers do not invade the spinal cord before stage 48, i.e., shortly after the lower limb buds arise. Reticulospinal and vestibulospinal fibers innervating the lumbar spinal cord, i.e., part of the definitive motor system, come from collaterals of early arising tail spinal cord-innervating reticulospinal and vestibulospinal axons, another part from a second wave of reticulospinal and vestibulospinal fibers. The ingrowth of rubrospinal fibers parallels the changes observed in locomotor pattern. Until stage 58, locomotion consists of coordinated, alternate contractions of the axial muscles on each side of the body. From stage 63 on, swimming is accomplished solely with the extremities. It is in this period that rubrospinal fibers innervate the lumbar spinal cord.

In amniotes, the ingrowth of reticulospinal, vestibulospinal and rubrospinal fibers into the spinal cord can also be correlated to the stage of development. At the time of appearance of the lower limb buds (HH17 in chickens, Carnegie stage 13 in rats and man), reticulospinal and interstitiospinal fibers reach the spinal cord, directly followed by vestibulospinal fibers. In chickens, rubrospinal fibers innervate the spinal cord by E7. In mammals, rubrospinal fibers do not innervate the spinal cord before the end of the embryonic period. Since in the North American opossum reticulospinal fibers already innervate the lumbar spinal cord at P1 and interstitiospinal and vestibulospinal fibers by P1–P3, it is likely that all these descending brainstem pathways reach the spinal cord and innervate the cervical enlargement prenatally. After birth, the forelimbs are used to climb into the mother's pouch. For such movements central motor control is necessary.

Although the corticospinal tract is always the latest to develop, clear differences are found in the time corticospinal fibers reach the spinal cord. Whereas in rodents corticospinal fibers reach the spinal cord at P0 or somewhat later, in the North American opossum the corticospinal tract does not reach spinal levels before P30, but in man pyramidal tract fibers extend already as far caudally as the pyramidal decussation at the end of the embryonic period. The first ingrowth of reticulospinal, interstitiospinal and vestibulospinal fibers can be clearly related to a particular developmental stage, but that of rubrospinal and corticospinal tracts can not (see Table 9). Nevertheless, the development of descending supraspinal pathways occurs asynchronously and according to a highly predictable sequence.

10.4
The Earliest Descending Supraspinal Fibers Arrive in a Rather Immature Spinal Cord

In the spinal cord of anurans, chickens and rodents, a sequential production of motor, relay, and interneuron populations has been shown. The earliest descending supraspinal fibers arrive in a rather immature spinal cord. In general, descending supraspinal, propriospinal, and ascending spinal projections appear to be formed around the same time with dorsal root projections clearly lagging behind. In *Xenopus laevis*, in the lumbar spinal cord the hindlimb-innervating motoneurons could be retrogradely labeled from the developing hindlimb bud at stage 48. At first these motoneurons bear only a few dorsal dendrites extending into the adjacent white matter. Already at stage 50 these dorsal dendrites have invaded the dorsolateral part of the marginal zone, and ventral dendrites are found. It seems likely that at this stage the first contacts with propriospinal and reticulospinal fibers are made. Dorsal root fibers exhibit a "waiting period" before entering the spinal gray matter at stage 53/54. Although the exact time of outgrowth of propriospinal and reticulospinal axons in the lumbar spinal cord of *X. laevis* is unknown, it is likely that propriospinal and reticulospinal fibers innervate the hindlimb-innervating motoneurons before muscle afferents do.

In chicken embryos, Oppenheim and co-workers showed that synapse formation from propriospinal sources, i.e., from spinal interneurons, precedes that from supraspinal descending axons (Oppenheim et al. 1988; Yaginuma et al. 1991, 1994). Sholomenko and O'Donovan (1995) showed that at E6 lumbosacral motor activity could be evoked by brainstem electrical stimulation, i.e., when the first detectable limb movements begin. The activation of the lumbosacral network for hindlimb movements is mediated by reticulospinal axons travelling in the ventral funiculus. Dorsal root fibers invade the gray matter of the lumbosacral spinal cord by E6, and extend into the vicinity of motoneuron dendrites by E7.5 (Davis et al. 1989b).

In rats, the earliest propriospinal and ascending spinal projections appear to be formed by E12, i.e., around the same time that motoneurons are developing and descending supraspinal projections start to invade the spinal cord. The earliest dorsal root projections enter the cervical spinal cord at E15 (Snider et al. 1992). At that stage of development, the lateral motoneurons extend their dendrites medially or dorsomedially into the direction of the incoming dorsal root fibers. It is likely that the first

Shapovalov AI (1972) Evolution of neuronal systems of suprasegmental motor control (in Russian). Neurophysiology (Kiev) 4:346–359

Shapovalov AI (1975) Neuronal organization and synaptic mechanisms of supraspinal motor control in vertebrates. Rev Physiol Biochem Pharmacol 72:1–54

Sharma SC, Jadhao AG, Prasada Rao PD (1993) Regeneration of supraspinal projection neurons in the adult goldfish. Brain Res 620:221–228

Shiga T, Oppenheim RW (1991) Immunolocalization studies of putative guidance molecules used by axons and growth cones of intersegmental interneurons in the chick embryo spinal cord. J Comp Neurol 310:234–252

Shiga T, Künzi R, Oppenheim RW (1991) Axonal projections and synaptogenesis by supraspinal descending neurons in the spinal cord of the chick embryo. J Comp Neurol 305:83–95

Shimizu I, Oppenheim RW, O'Brien M, Shneiderman A (1990) Anatomical and functional recovery following spinal cord transection in the chick embryo. J Neurobiol 21:918–937

Sholomenko GN, Delaney KR (1998) Restitution of functional neural connections in chick embryos assessed in vitro after spinal cord transection in ovo. Exp Neurol 154:430–451

Sholomenko GN, O'Donovan MJ (1995) Development and characterization of pathways descending to the spinal cord in the embryonic chick. J Neurophysiol 73:1223–1233

Sholomenko GN, Steeves JD (1987) Effects of selective spinal cord lesions on hind limb locomotion in birds. Exp Neurol 95:403–418

Shumway W (1940) Stages in the normal development of Rana pipiens. I. External form. Anat Rec 78:139–144

Sillar KT (1991) Spinal pattern generation and sensory gating mechanisms. Curr Opin Neurobiol 1:583–589

Sillar KT, Roberts A (1988) A neuronal mechanism for sensory gating during locomotion in a vertebrate. Nature 331:262–265

Sillar KT, Roberts A (1992) The role of premotor interneurons in phase-dependent modulation of a cutaneous reflex during swimming in Xenopus laevis embryos. J Neurosci 12:1647–1657

Sillar KT, Wedderburn JF, Simmers AJ (1992) Modulation of swimming rhythmicity by 5-hydroxytryptamine during post-embryonic development in Xenopus laevis. Proc Roy Soc Lond (Biol) 250:107–114

Sillar KT, Woolston AM, Wedderburn JF (1995) Involvement of brainstem serotonergic interneurons in the development of a vertebrate spinal locomotor circuit. Proc Roy Soc Lond (Biol) 259:65–70

Silos-Santiago I, Snider WD (1992) Development of commissural neurons in the embryonic rat spinal cord. J Comp Neurol 325:514–526

Silos-Santiago I, Snider WD (1994) Development of interneurons with ipsilateral projections in embryonic rat spinal cord. J Comp Neurol 342:221–231

Silver J, Sidman RL (1980) A mechanism for the guidance and topographic patterning of retinal ganglion cell axons. J Comp Neurol 189:101–111

Silver J, Lorenz SE, Wahlsten D, Coughlin J (1982) Axonal guidance during development of the great cerebral commissures: Descriptive and experimental studies, in vivo, on the role of preformed glial pathways. J Comp Neurol 210:10–29

Silver J, Edwards MA, Levitt P (1993) Immunocytochemical demonstration of early appearing astroglial structures that form boundaries and pathways along axon tracts in the fetal brain. J Comp Neurol 328:415–436

Simpson SB Jr (1983) Fasciculation and guidance of regenerating central axons by the ependyma. In: Kao CC, Bunge RP, Reier PJ (eds) Spinal Cord Reconstruction. Raven Press, New York, pp 151–162

Sims RT (1962) Transection of the spinal cord in developing Xenopus laevis. J Embryol Exp Morphol 10:115–126

Sims TJ (1977) The development of monoamine-containing neurons in the brain and spinal cord of the salamander, Ambystoma mexicanum. J Comp Neurol 173:319–336

Sims TJ, Vaughn JE (1979) The generation of neurons involved in an early reflex pathway of embryonic rat spinal cord. J Comp Neurol 183:707–720

Singer M, Caston JD (1972) Neurotrophic dependence of macromolecular synthesis in the early limb regenerate of the newt, Triturus. J Embryol Exp Morphol 28:1–11

Singer M, Nordlander RH, Egar M (1979) Axonal guidance during embryogenesis and regeneration in the spinal cord of the newt: The blueprint hypothesis of neuronal pathway patterning. J Comp Neurol 185:1–22

Sive HL, Cheng PF (1991) Retinoic acid perturbs the expression of Xhox-lab genes and alters mesodermal determination in *Xenopus laevis*. Genes Dev 5:1321–1332

Smith CL, Frank E (1988a) Specificity of sensory projections in the spinal cord during development in bullfrogs. J Comp Neurol 269:96–108

Smith CL, Frank E (1988b) Peripheral specification of sensory connections in the spinal cord. Brain Behav Evol 31:227–242

Snider WD, Zhang L, Yusoof S, Gorukanti N, Tsering C (1992) Interactions between dorsal root axons and their target motor neurons in developing mammalian spinal cord. J Neurosci 12:3494–3508

Soffe SR (1987) Ionic and pharmacological properties of reciprocal inhibition in Xenopus embryo motoneurons. J Physiol (Lond) 382:463–473

Soffe SR, Roberts A (1982a) The activity of myotomal motoneurons during fictive swimming in frog embryos. J Neurophysiol 48:1274–1278

Soffe SR, Roberts A (1982b) Tonic and phasic synaptic inputs to spinal cord motoneurons active during fictive locomotion in frog embryos. J Neurophysiol 48:1279–1288

Soffe SR, Clarke JDW, Roberts A (1984) Activity of commissural interneurons in spinal cord of Xenopus embryos. J Neurophysiol 51:1257–1267

Song W-J, Murakami F (1998) Development of functional topography in the corticorubral projection: An *in vivo* assessment using synaptic potentials recorded from fetal and newborn cats. J Neurosci 18:9354–9364

Song W-J, Kanda M, Murakami F (1995a) Prenatal development of cerebrorubral and cerebellorubral projections in cats. Neurosci Lett 200:41–44

Song W-J, Okawa K, Kanda M, Murakami F (1995b) Perinatal development of action potential propagation in cat rubrospinal axons. J Physiol (Lond) 488:419–426

Stanfield BB (1992) The development of the corticospinal projection. Prog Neurobiol 38:169–202

Stanfield BB, O'Leary DDM (1985) The transient corticospinal projection from the occipital cortex during the postnatal development of the rat. J Comp Neurol 238:236–248

Steeves JD, Sholomenko GN, Webster DMS (1987) Stimulation of the pontomedullary reticular formation initiates locomotion in decerebrate birds. Brain Res 401:205–212

Steeves JD, Keirstead HS, Ethell DW, Hasan SJ, Muir GD, Pataky DM, McBride CB, Petrausch B, Zwimpfer TJ (1994) Permissive and restrictive periods for brainstem-spinal regeneration in the chick. Prog Brain Res 103:243–262

Stehouwer DJ (1986) Behavior of larval and juvenile bullfrogs (*Rana catesbeiana*) following chronic spinal cord transection. Behav Neural Biol 45:120–134

Stehouwer DJ (1992) Development of anuran locomotion: Ethological and neurophysiological considerations. J Neurobiol 23:1467–1485

Stehouwer DJ, Farel PB (1984) Development of hindlimb locomotor behavior in the frog. Dev Psychol 17:217–232

Steinbusch HWM (1981) Distribution of serotonin-immunoreactivity in the central nervous system of the rat – cell bodies and terminals. Neuroscience 6:557–618

Steinbusch HWM, Verhofstad AAJ, Joosten HWJ (1983) Antibodies to serotonin for neuroimmunocytochemical studies on the central nervous system. In: Cuello C (ed) Neuroimmunocytochemistry. IBRO Handbook Series Methods in the Neurosciences, Vol 3. Wiley, New York, pp 193–214

Stensaas LJ (1983) Regeneration in the spinal cord of the newt Notophthalmus (Triturus) pyrrhogaster. In: Kao CC, Bunge RP, Reier PJ (eds) Spinal Cord Reconstruction. Raven Press, New York, pp 121–149

Streeter GL (1951) Developmental Horizons in Human Embryos. Age group XI to XXIII. Collected papers from the Contributions to Embryology. Embryol Reprint Vol II. Carnegie Instn Washington, Washington, DC

Stuermer CAO, Bastmeyer M, Bähr M, Strobel G, Paschke K (1992) Trying to understand axonal regeneration in the CNS of fish. J Neurobiol 23:537–550

Szaro BG, Lee VM-Y, Gainer H (1989) Spatial and temporal expression of phosphorylated and non-phosphorylated forms of neurofilament proteins in the developing nervous system of *Xenopus laevis*. Dev Brain Res 48:87–103

Szaro BG, Grant P, Lee VM-Y, Gainer H (1991) Inhibition of axonal development after injection of neurofilament antibodies into a *Xenopus laevis* embryo. J Comp Neurol 308:576–585

Tan H, Miletic V (1990) Bulbospinal serotininergic pathways in the frog Rana pipiens. J Comp Neurol 292:291–302

Tank PW, Holder N (1981) Pattern regulation in the limbs of urodelean amphibians. Ann Rev Biol 56:113–142

Taylor AC, Kollros JJ (1946) Stages in the normal development of Rana pipiens larvae. Anat Rec 94:7–23

Tello F (1923) Les différenciations neuronales dans l'embryon du poulet, pendant les premiers jours de l'incubation. Trav Lab Rech Biol Univ Madrid 21:1–93

ten Donkelaar HJ (1982) Organization of descending pathways to the spinal cord in amphibians and reptiles. Prog Brain Res 57:25–67

ten Donkelaar HJ (1988) Evolution of the red nucleus and rubrospinal tract. Behav Brain Res 28:9–20

ten Donkelaar HJ (1990) Brainstem mechanisms of behavior. In: Klemm WR, Vertes RP (eds) Brainstem Mechanisms in Behavior. Wiley, New York, pp 199–237

ten Donkelaar HJ (1994) Some notes on the organization of spinal and supraspinal premotor networks for locomotion. Eur J Morphol 32:156–167

ten Donkelaar HJ (1998a) Urodeles. In: Nieuwenhuys R, ten Donkelaar HJ, Nicholson C The Central Nervous System of Vertebrates. Springer, Berlin-Heidelberg-New York, pp 1045–1150

ten Donkelaar HJ (1998b) Anurans. In: Nieuwenhuys R, ten Donkelaar HJ, Nicholson C The Central Nervous System of Vertebrates. Springer, Berlin-Heidelberg-New York, pp 1151–1314

ten Donkelaar HJ (1999) Evolution of vertebrate motor systems. In: Roth G, Wullimann M (eds) Brain Evolution and Cognition. Spektrum/Wiley, Heidelberg, in press

ten Donkelaar HJ, de Boer-van Huizen R (1982) Observations on the development of descending pathways from the brain stem to the spinal cord in the clawed toad, *Xenopus laevis*. Anat Embryol 163:461–473

ten Donkelaar HJ, de Boer-van Huizen R (1991) Observations on the development of ascending spinal pathways in the clawed toad, *Xenopus laevis*. Anat Embryol 183:589–603

ten Donkelaar HJ, Nicholson C (1998) Notes on techniques. In: Nieuwenhuys R, ten Donkelaar HJ, Nicholson C The Central Nervous System of Vertebrates. Springer, Berlin-Heidelberg-New York, pp 327–355

ten Donkelaar HJ, Geysberts LGM, Dederen PJW (1979) Stages in the prenatal development of the Chinese hamster (*Cricetulus griseus*). Anat Embryol 156:1–28

ten Donkelaar HJ, Kusuma A, de Boer-van Huizen R (1980) Cells of origin of pathways descending to the spinal cord in some quadrupedal reptiles. J Comp Neurol 192:827–851

ten Donkelaar HJ, de Boer-van Huizen R, Schouten FTM, Eggen SJH (1981) Cells of origin of descending pathways to the spinal cord in the clawed toad (*Xenopus laevis*). Neuroscience 6:2297–2312

ten Donkelaar HJ, Bangma GC, Barbas-Henry HA, de Boer-van Huizen R, Wolters JG (1987) The Brain Stem in a Lizard, *Varanus exanthematicus*. Adv Anat Embryol Cell Biol, Vol 107

ten Donkelaar HJ, de Boer-van Huizen R, van der Linden JAM (1991) Early development of rubrospinal and cerebellorubral projections in *Xenopus laevis*. Dev Brain Res 58:297–300

ten Donkelaar HJ, de Boer-van Huizen R, Bergervoet-Vernooy I (1993) Development, plasticity and regeneration of reticulospinal pathways: A case study in *Xenopus laevis*. Soc Neurosci Abst 19:617

ten Donkelaar HJ, Wesseling P, Semmekrot BA, Liem KD, Tuerlings J, Cruysberg JRM, de Wit PEJ (1999) Severe, non X-linked congenital microcephaly with absence of the pyramidal tracts in two siblings. Acta Neuropathol 98:203–211

Terashima T (1995a) Anatomy, development and lesion-induced plasticity of rodent corticospinal tract. Neurosci Res 22:139–161

Terashima T (1995b) Course and collaterals of corticospinal fibers arising from the sensorimotor cortex in the reeler mouse. Dev Neurosci 17:8–19

Terashima T, Inoue K, Inoue Y, Mikoshiba K, Tsukada Y (1983) Distribution and morphology of corticospinal tract neurons in reeler mouse cortex by the retrograde HRP method. J Comp Neurol 218:314–326

Terman JR, Wang XM, Martin GF (1996) Growth of dorsal spinocerebellar axons through a lesion of their spinal pathway during early development in the North American opossum, *Didelphis virginiana*. Dev Brain Res 93:33–48

Terman JR, Wang XM, Martin GF (1997) Developmental plasticity of selected spinocerebellar axons. Studies using the North American opossum, *Didelphis virginiana*. Dev Brain Res 102:309–314

Terman JR, Wang XM, Martin GF (1999) Developmental plasticity of ascending spinal axons. Studies using the North American opossum, *Didelphis virginiana*. Dev Brain Res 112:65–77

Tessier-Lavigne M, Goodman CS (1996) The molecular biology of axon guidance. Science 274:1123–1132

Teztlaff W, Kobayashi NR, Giehl KMG, Tsui BJ, Cassar SL, Bedard AM (1994) Response of rubrospinal and corticospinal neurons to injury and neurotrophins. Prog Brain Res 103:271–286

Theiler K (1972) The House Mouse. Development and normal stages from fertilization to 4 weeks of age. Springer, Berlin-Heidelberg

Theriault E, Tatton WG (1989) Postnatal redistribution of pericruciate motor cortical projections within the kitten spinal cord. Dev Brain Res 45:219–237

Tohyama M, Sakai K, Salvert D, Touret M, Jouvet M (1979a) Spinal projections from the lower brain stem in the cat as demonstrated by the HRP technique. I. Origins of the reticulospinal tracts and their funicular trajectories. Brain Res 173:383–403

Tohyama M, Sakai K, Salvert D, Touret M, Jouvet M (1979b) Spinal projections from the lower brain stem in the cat as demonstrated by the HRP technique. II. Projections from the dorsal pontine tegmentum and raphe nuclei. Brain Res 176:215–231

Tohyama T, Lee VM-Y, Rorke LB, Trojanowski JQ (1991) Molecular milestones that signal axonal maturation and the commitment of human spinal cord precursor cells to the neuronal or glial phenotype in development. J Comp Neurol 310:285–299

Torvik A, Brodal A (1957) The origin of reticulospinal fibers in the cat. An experimental study. Anat Rec 128:113–137

Tóth P, Csank G, Lázár G (1985) Morphology of the cells of origin of descending pathways to the spinal cord in *Rana esculenta*. J Hirnforsch 26:365–383

Treherne JM, Woodward SKA, Varga ZM, Ritchie JM, Nicholls JG (1992) Restoration of conduction and growth of axons through injured spinal cord of neonatal opossum in culture. Proc Nat Acad Sci USA 89:431–434

Tyndale-Biscoe CH, Janssens PA, eds (1987) The Developing Marsupial. Models for biomedical research. Springer, New York

Ulinski PS (1971) External morphology of pouch young opossum brains: A profile of opossum neurogenesis. J Comp Neurol 142:33–58

Vaage S (1969) The Segmentation of the Primitive Neural Tube in Chick Embryos (Gallus domesticus). Adv Anat Embryol Cell Biol 41:1–88

van der Linden JAM, ten Donkelaar HJ (1987) Observations on the development of cerebellar afferents in *Xenopus laevis*. Anat Embryol 176:431–439

van der Linden JAM, ten Donkelaar HJ (1990) Morphological evidence for a monosynaptic connection between cerebellar Purkinje cells and vestibulospinal tract neurons in the larval clawed toad, *Xenopus laevis*. Neurosci Lett 112:121–126

van der Linden JAM, ten Donkelaar HJ, de Boer-van Huizen R (1988) Development of spinocerebellar afferents in the clawed toad, *Xenopus laevis*. J Comp Neurol 277:41–52

van Mier P (1986) The Development of the Motor System in the Clawed Toad, Xenopus laevis. Thesis, University of Nijmegen

van Mier P, ten Donkelaar HJ (1984) Early development of descending pathways from the brain stem to the spinal cord in *Xenopus laevis*. Anat Embryol 170:295–306

van Mier P, ten Donkelaar HJ (1988) The development of primary afferents to the lumbar spinal cord in *Xenopus laevis*. Neurosci Lett 84:35–40

van Mier P, ten Donkelaar HJ (1989) Structural and functional properties of the reticulospinal neurons in the early swimming stage *Xenopus* embryo. J Neurosci 9:25–37 van Mier P, van Rheden R, ten Donkelaar HJ (1985) The development of the dendritic organization of primary and secondary motoneurons: An HRP study in *Xenopus laevis*. Anat Embryol 172:311–324

van Mier P, Joosten HWJ, van Rheden R, ten Donkelaar HJ (1986) The development of serotonergic raphespinal projections in *Xenopus laevis*. Int J Devl Neurosci 4:465–475

van Mier P, Armstrong J, Roberts A (1989) Development of early swimming in *Xenopus laevis* embryos: Myotomal musculature, its innervation and activation. Neuroscience 32:113–126

Varga ZM, Bandtlow CE, Erulkar SD, Schwab ME, Nicholls JG (1995a) The critical period for repair of CNS of neonatal opossum (*Monodelphis domestica*) in culture: Correlation with development of glial cells, myelin and growth-inhibiting molecules. Eur J Neurosci 7:2119–2129

Varga ZM, Schwab ME, Nicholls JG (1995b) Myelin-associated neurite growth-inhibiting proteins and suppression of regeneration of immature mammalian spinal cord in culture. Proc Natl Acad Sci USA 92:10959–10963

Vargas-Lizardi P, Lyser K (1974) Time of origin of Mauthner's neurons in Xenopus laevis embryos. Dev Biol 38:220–228

Vaughn JE, Sims T, Nakashima M (1977) A comparison of the early development of axodendritic and axosomatic synapses upon embryonic mouse spinal motor neurons. J Comp Neurol 175:79–100

Verhaart WJC, Kramer W (1952) Uncrossed pyramidal tract. Acta Psychiat Neurol Scand 27:181–200

Verney C, Zecevic N, Nikolic B, Alvarez C, Berger B (1991) Early evidence of catecholaminergic cell groups in 5- and 6-week-old human embryos using tyrosine hydroxylase and dopamine-ß-hydroxylase immunocytochemistry. Neurosci Lett 131:121–124

von Meyenburg J, Brösamle C, Metz GAS, Schwab ME (1998) Regeneration and sprouting of chronically injured corticospinal tract fibers in adult rats promoted by NT-3 and the mAb IN-1, which neutralizes myelin-associated neurite growth inhibitors. Exp Neurol 154:583–594

Wang XM, Xu XM, Qin YQ, Martin GF (1992) The origins of supraspinal projections to the cervical and lumbar spinal cord at different stages of development in the gray short-tailed Brazilian opossum, *Monodelphis domestica*. Dev Brain Res 68:203–216

Wang XM, Qin YQ, Xu XM, Martin GF (1994) Developmental plasticity of reticulospinal and vestibulospinal axons in the North American opossum, *Didelphis virginiana*. J Comp Neurol 349:288–302

Wang XM, Terman JR, Martin GF (1996) Evidence for growth of supraspinal axons through the lesion after transection of the thoracic spinal cord in the developing opossum, *Didelphis virginiana*. J Comp Neurol 371:104–115

Wang XM, Qin YQ, Terman JR, Martin GF (1997) Early development and developmental plasticity of the fasciculus gracilis in the North American opossum (*Didelphis virginiana*). Dev Brain Res 98:151–163

Wang XM, Terman JR, Martin GF (1998a) Regeneration of supraspinal axons after transection of the thoracic spinal cord in the developing opossum, *Didelphis virginiana*. J Comp Neurol 398:83–97

Wang XM, Basso DM, Terman JR, Bresnahan JC, Martin GF (1998b) Adult opossums (*Didelphis virginiana*) demonstrate near normal locomotion after spinal cord transection as neonates. Exp Neurol 151:50–69

Webster DMS, Steeves JD (1988) Origins of brainstem-spinal projections in the duck and goose. J Comp Neurol 273:573–583

Webster DMS, Rogers LJ, Pettigrew JD, Steeves JD (1990) Origins of descending spinal pathways in prehensile birds: Do parrots have a homologue to the corticospinal tract of mammals? Brain Behav Evol 36:216–226

Weidenheim KM, Kress Y, Epshteyn I, Rashbaum WK, Lyman WD (1992) Early myelination in the human fetal lumbosacral spinal cord: Characterization by light and electron microscopy. J Neuropathol Exp Neurol 51:142–149

Weidenheim KM, Epshteyn I, Rashbaum WK, Lyman WD (1993) Neuroanatomical localization of myelin basic protein in the late first and early second trimester human foetal spinal cord and brainstem. J Neurocytol 22:507–516

Weidenheim KM, Bodhireddy SR, Rashbaum WK, Lyman WD (1996) Temporal and spatial expression of major myelin proteins in the human fetal spinal cord during the second trimester. J Neuropathol Exp Neurol 55:734–745

Weikert T, Rathjen FG, Layer PG (1990) Developmental maps of acetylcholinesterase and G4-antigen of the early chicken embryo: Long-distance tracts originate from AChE-producing cell bodies. J Neurobiol 21:482–498

Wentworth LE (1984a) The development of the cervical spinal cord of the mouse embryo. I. A Golgi analysis of ventral root neuron differentiation. J Comp Neurol 222:81–95

Wentworth LE (1984b) The development of the cervical spinal cord of the mouse embryo. II. A Golgi analysis of sensory, commissural, and association cell differentiation. J Comp Neurol 222:96–115

Wessels WJT, Feirabend HKP, Marani E (1991) Development of projections of primary afferent fibers from the hindlimb to the gracile nucleus: A WGA-HRP study in the rat. Dev Brain Res 63:265–279

Westerga J, Gramsbergen A (1990) The development of locomotion in the rat. Dev Brain Res 57:163–174

Whishaw IQ, Kolb B (1988) Sparing of skilled forelimb reaching and corticospinal projections after neonatal motor cortex removal or hemidecortication in the rat: Support for the Kennard doctrine. Brain Res 451:97–114

Will U (1986) Mauthner cells survive metamorphosis in anurans: A comparative HRP study on the cytoarchitecture of Mauthner neurons in amphibians. J Comp Neurol 244:111–120

Will U (1991) Amphibian Mauthner cells. Brain Behav Evol 37:31–332

Wilson JG (1973) Environment and Birth Defects. Academic Press, New York

Wilson SW, Ross LS, Parrett T, Easter SS (1990) The development of a simple scaffold of axon tracts in the brain of the embryonic zebrafish, *Brachydanio rerio*. Development 108:121–145

Wilson SW, Brennan C, Macdonald R, Brand M, Holder N (1997) Analysis of axon tract formation in the zebrafish brain: The role of territories of gene expression and their boundaries. Cell Tissue Res 290:189–196

Windle WF (1932a) The neurofibrillar structure of the 7-mm cat embryo. J Comp Neurol 55:99–138

Windle WF (1932b) The neurofibrillary structure of the five-and-one-half-millimeter cat embryo. J Comp Neurol 55:315–331

Windle WF (1935) Neurofibrillar development of cat embryos: Extent of development in the telencephalon and diencephalon up to 15 mm. J Comp Neurol 63:139–171

Windle WF (1970) Development of neural elements in human embryos of four to seven weeks gestation. Exp Neurol, Suppl 5:44–83

Windle WF, Austin MF (1936) Neurofibrillar development in the central nervous system of chick embryos up to 5 days incubation. J Comp Neurol 63:431–463

Windle WF, Baxter RE (1936) The first neurofibrillar development in albino rat embryos. J Comp Neurol 63:173–199

Windle WF, Fitzgerald JE (1937) Development of the spinal reflex mechanism in human embryos. J Comp Neurol 67:493–509

Windle WF, Fitzgerald JE (1942) Development of the human mesencephalic trigeminal root and related neurons. J Comp Neurol 77:597–608

Windle WF, Griffin AM (1931) Observations on embryonic and fetal movements of the cat. J Comp Neurol 52:159–188

Wirth FP, O'Leary JL, Smith JM, Jenny AB (1974) Monosynaptic corticospinal-motoneuron path in the raccoon. Brain Res 77:344–348

Wise SP, Hendry SHC, Jones EG (1977) Prenatal development of sensorimotor cortical projections in cats. Brain Res 138:538–544

Witschi E (1972) In: Altman PL, Dittmer DS (eds) Biology Data Book, Vol I. Fed Amer Soc Exp Biology, Bethesda, pp 176–180

Woodward SKA, Treherne JM, Knott GW, Fernandez J, Varga ZM, Nicholls JG (1993) Development of connections by axons growing through injured spinal cord of neonatal opossum in culture. J Exp Biol 176:77–88

Woźniak W, O'Rahilly R (1982) An electron microscopic study of myelination of pyramidal fibers at the level of the pyramidal decussation. J Hirnforsch 23:331–342

Xu XM, Martin GF (1989) Developmental plasticity of the rubrospinal tract in the North American opossum. J Comp Neurol 279:368–381

Xu XM, Martin GF (1991) Evidence for new growth and regeneration of cut axons in developmental plasticity of the rubrospinal tract in the North American opossum. J Comp Neurol 313:103–112

Xu XM, Martin GF (1992) The response of rubrospinal neurons to axotomy at different stages of development in the North American opossum. J Neurotrauma 9:93–105

Yaginuma H, Homma S, Künzi R, Oppenheim RW (1991) Pathfinding by growth cones of commissural interneurons in the chick embryo spinal cord: A light and electron microscopic study. J Comp Neurol 304:78–102

Yaginuma H, Shiga T, Oppenheim RW (1994) Early developmental patterns and mechanisms of axonal guidance of spinal interneurons in the chick embryo spinal cord. Prog Neurobiol 44:249–278

Yakovlev PI, Lecours AR (1967) The myelogenetic cycles of regional maturation of the brain. In: Minkowski A (ed) Regional Development of the Brain In Early Life. Blackwell, Oxford, pp 3–70

Yakovlev PI, Rakic P (1966) Patterns of decussation of bulbar pyramids and distribution of pyramidal tracts on two sides of the spinal cord. Trans Am Neurol Assoc 91:366–367

Yin HS, Selzer ME (1983) Axonal regeneration in lamprey spinal cord. J Neurosci 3:1135–1144

Zamora AJ, Mutin M (1988) Vimentin and glial fibrillary acidic protein filaments in radial glia of the adult urodele spinal cord. Neuroscience 27:279–288

Zecevic N, Verney C (1995) Development of the catecholamine neurons in human embryos and fetuses, with special emphasis on the innervation of the cerebral cortex. J Comp Neurol 351:509–535

Z'Graggen WJ, Metz GAS, Kartje GL, Thallmair M, Schwab ME (1998) Functional recovery and enhanced corticofugal plasticity after unilateral pyramidal tract lesion and blockade of myelin-associated neurite growth inhibitors in adult rats. J Neurosci 18:4744–4757

Zottoli SJ, Bentley AP, Feiner DG, Hering JR, Prendergast BJ, Rieff HI (1994) Spinal cord regeneration in adult goldfish: Implications for functional recovery in vertebrates. Prog Brain Res 103:219–228

Subject Index

144

Printing and binding: Konrad Triltsch, Print und digitale Medien, 97070 Würzburg